Radar Targets

Radar Targets

Christian G. Bachman
KRYBOK Associates

LexingtonBooks
D.C. Heath and Company
Lexington, Massachusetts
Toronto

Library of Congress Cataloging in Publication Data

Bachman, Christian G., 1931-
Radar targets.
Includes index.
1. Radar targets. I. Title.
TK6585.B24 621.3848'1 81-48003
ISBN 0-669-05232-9 AACR2

Published simultaneously in Canada.

Printed in the United States of America.

International Standard Book Number: 0-669-05232-9

Library of Congress Catalog Card Number: 81-48003

Contents

List of Figures and Tables

Figures

Tables

Acknowledgments

Professor Andrew L. Maffett of the University of Michigan provided his notes on radar cross-section estimation and control for incorporation in this book. His contributions, as well as those of the late Professor Keeve M. Siegel and John C. Crispin, Jr., are gratefully acknowledged.

Fellow authors Howard E. King of the Aerospace Corporation and Dr. Robert C. Hansen, now of R.C. Hansen, Inc., contributed to the material on radar cross-section measurements.

Dr. Daniel E. Levine of Lockheed Missiles and Space Company pioneered the use of cartographic techniques in prediction of spatial-coverage efficacy of radar-enhancement devices, such as corner reflectors.

Professional colleagues Robert E. Kell and Marley E. Bechtel, formerly of Cornell Aeronautical Laboratory (now CALSPAN Corporation), made fundamental contributions to the material and to my understanding of radar-target identification, as did W. Philip Melling, the late Dr. Gordon A.R. Graham, and Professor Robert D. Stratton of the University of Oklahoma.

Dr. Hillard W. Wachowski, a former fellow employee at the Aerospace Corporation, provided the material on radar propagation effects, based largely on internal reports at General Electric Company by Dr. George H. Millman.

This book is dedicated to Arnold E. Resnik, mentor, colleague, and long-time friend, and to Dr. Vernal Josephson of the Aerospace Corporation, who suggested it be written.

1 Radar Cross-Section Estimation and Control

Introduction and Summary

There are two ways to find the radar-reflectivity characteristics of a target: theoretical calculation and experimental measurement (both static and dynamic). Neither functions perfectly without the other. Either, however, when properly augmented by the other, can yield a good description of a target signature.

Both approaches are complicated by the large number of parameters on which the radar cross-section (RCS) of a scatterer depends: the shape, size, and material constituency of the target in relation to the frequency, polarizations, and aspects of illumination and observation. In particular, if illumination and observation points are at separated locations (*bistatic operation*) rather than at essentially the same location (*monostatic operation*), the number of parameters describing polarization and aspect is doubled. If then the target is imagined to be fixed at the center of a very large sphere, RCS can be viewed as a (pattern) function of target characteristics measured in wavelengths defined for every possible location of illumination and observation points, and polarization directions at these points, on the surface of the sphere. The large-sphere radius is usually dictated by physical situations of interest, not by limitations inherent to the scattering problem. However, this large radius does make possible the advantage of being able to view both incident and reflected wavefronts as plane; and these are usually more easily dealt with, both theoretically and experimentally, than nonplanar wavefronts (additionally, any wave can be represented as a superposition of plane waves).

An exact solution to the scattering problem for a complex target is out of the question; in fact, exact solutions exist only for a few simple shapes. Acceptable estimates based on approximation methods are available, however, and are relatively straightforward in their application. What should be taken as a definition of *acceptable*? RCS analysts may disagree on this term, but generally speaking, if theoretical estimate and experimental measurement can be made to agree to within 3–5 dB, then the agreement is said to be acceptable. This criterion places a requirement on measurement techniques as well as on theoretical estimation methods. If, for example, the experiment is dynamic, then there is an aspect uncertainty that makes impossible an exact duplication of any given measurement. This fact is particularly true of a very complex scatterer measured at a wavelength that is small compared with the dimensions of most of the individual scattering components. In this type of situation, pattern peaks and nulls will be extremely dependent on relative phases among scattering centers, so that even small variations of aspect will cause large changes in scattering amplitudes. It is doubtless for this reason that experimenters choose to process such data by recording medians or maxima over, say, 10° aspect intervals.

There are also difficulties with static measurements. Background noise is especially troublesome in the measurement of low-RCS targets. Hence great care must be taken to minimize the RCS of other scatterers present in the environment of the target. One of these in particular is the device by which the target is supported. Indeed, the target support may play a significantly restrictive role for other than just low-RCS targets. For example, if the target must be large and heavy so that it can be manipulated into only a limited number of orientations, then the aspect intervals at which RCS can be measured will be correspondingly restricted.

There would seem then to be no panacea for RCS determination realizable by theoretical calculations or by experimental measurements (either static or dynamic). However, theoretical work has guided and interpreted many measurement methods. Measurements have helped immensely to spur and confirm many theoretical treatments, particularly for simple shapes and for complex targets composed of such simple shapes. In the next section I shall discuss, for the monostatic case, the state of these methods and the kinds of results they produce, and give examples of the agreement these results can be expected to have with measured data. In the third section I shall offer a similar discussion of the state of the art for the bistatic situation.

For a great many applications, certain systems requirements may dictate target configurations that produce undesired RCS signatures. In such cases the RCS may be modified (usually reduced, although enhancement is sometimes required) by changing the shape or the material constituency of the target, often without encroaching unduly on the original systems requirements. With shape modifications, the intent is to direct the RCS return from crucial to less sensitive aspect regions or frequency domains. With material-constituency modification, the reflective characteristics of the target are changed by changing its surface impedance (that is, its permittivity and/or permeability); such modifications may be structural (for example, fiberglass wings or fins) or may be introduced by coating metallic surfaces with a suitable radar-absorbing material (RAM). These matters will be discussed further in the fourth section.

The final section of this chapter is a brief discussion of the camouflage of antennas and sensors, including possible treatments that can be used for radomes and canopies. In addition, I will consider briefly the subject of impedance loading, which is a variation of the modification of a target surface with RAM, and its application to long, thin bodies in particular.

RCS Estimation Methods

The RCS of a complex target can be described statistically on the basis of an equally likely relative phase relation among any two of the collection of simple scattering centers, or components, of which the complex target is made up [1,2]. Briefly, the idea is this: Suppose that the mth of M scattering centers has a RCS σ_m and a relative phase ϕ_m, where the σ_m are functions of aspect, frequency, polarization, and the constituent parameters of the scattering component. Suppose further that the ϕ_m are randomly, and independently, distributed between 0 and 2π—that is, that in the interval $0 \leq \phi_m \leq 2\pi$, the probability density $p(\phi_m) = 1/2\pi$, $m = 1, \ldots, M$. Then the average RCS $E(\sigma)$ over the M phase quantities ϕ_m is found to be

$$E(\sigma) = \sum_{m=1}^{M} \sigma_m, \tag{1.1}$$

and the rms spread about the average is given by

$$E(\sigma) \pm S, \tag{1.2}$$

where the deviation S is determined from

$$S^2 = \left(\sum_{m=1}^{M} \sigma_m \right)^2 - \sum_{m=1}^{M} \sigma^2{}_m = [E(\sigma)]^2 - \sum_{m=1}^{M} \sigma^2{}_m. \tag{1.3}$$

This result is often referred to as the *random phase* estimate of RCS for a complex target.

There has been no demonstrated need for statistics of an order higher than those obtainable from $E(\sigma)$ and $E(\sigma^2)$. However, it is often useful to have an upper bound to possible RCS values for a complex target, and this can be had from

$$\sigma_{\max} = \left| \sum_{m=1}^{M} \sqrt{\sigma_m} \right|^2. \tag{1.4}$$

We shall see that equations 1.1 through 1.4 furnish a perfectly adequate basis for theoretical estimates of complex-target RCS in the sense that such estimates agree well with experimentally measured data. Indeed, experience in comparing this type of estimate with experimentally measured data has shown that even a rapidly varying experimental RCS curve will lie within the region determined by the rms spread 50–60 percent of the time.

If the relative phases ϕ_m can be accurately determined, then a correspondingly more accurate estimate of RCS can be had from the so-called relative-phase result given by

$$\sigma_p = \left| \sum_{m=1}^{M} \sqrt{\sigma_m} \exp(i\phi_m) \right|^2. \tag{1.5}$$

Since ϕ_m is proportional to a distance d_m (from some arbitrary phase center to a scattering center) measured in wavelengths λ, if $d_{mm} \gg \lambda$, then it is very difficult to determine d_m/λ with sufficient accuracy to warrant the use of equation 1.5. It is usually for this reason that the random-phase method (equations 1.1–1.4) is to be preferred over the relative-phase method (equation 1.5). If, however, the quantities d_m/λ can be measured with desired accuracy, then σ_p will provide an RCS estimate for a complex target that is as good as the accuracy of the σ_m for the individual scattering centers. Mutual interactions among the separate scattering components are, except for those indicated by the relative phases ϕ_m, essentially negligible.

This brings us then to the component, or simple, shapes themselves. Work in this area will surely continue as long as there exists interest in radar reflectivity.

We must first try to establish a definition for *simple shape.* Although there is far from universal agreement on what represents a simple shape, it might suffice to say that a simple shape is characterized by having just one dominant type of scattering mechanism. Alternatively, we might choose the set of simple shapes to be those representable as coordinate surfaces in an orthogonal coordinate system and for which the problem of plane-wave scattering is exactly soluble. The first, though perhaps less exact, is more useful since it includes such desirable (for purposes of analyzing a complex target) basic bodies as the finite cone, the finite cylinder, and the finite flat plate. The first overlaps the second in that it contains such additional—and no less desirable—basic bodies as the sphere, the wedge, the torus, and so forth. More will be said about a simple-shape definition after certain approximation methods have been discussed.

One might think that if the class of simple shapes suggested by the second definition were extensive enough, the complex-body problem would become cut and dried, but this is not so. The exact solutions are as a rule simply too cumbersome to be used conveniently to solve a complex-body RCS problem. Indeed, the complex body is usually constituted of simple shapes in such a way as to make it impossible to use their exact solutions. A simple case in point is the wedge-cylinder (that is, a wedge with sides tangent to a cylinder), wherein the simple shapes (wedge and cylinder) are joined in such a way as to nullify the use of either exact solution. However, exact solutions for simple bodies are very useful in evaluating approximate solutions. It is these approximate solutions, very often asymptotic for large or small wavelength and usually interpretable in some significant physical sense, on which is built an RCS estimate for a complex target.

We now turn to a brief discussion of some of the commonly used approximate-solution techniques. I will classify these according to their frequency region of applicability—optics, Rayleigh, and resonance—which are determined by whether a major dimension is, respectively, much larger than, much smaller than, or comparable to a wavelength.

Optics Methods

The most commonly used optics methods for RCS determination are those that lend their name to the short-wavelength region, namely geometric optics and physical optics. Detailed discussion of these methods may be found in [2]. Briefly, the geometric-optics approach is based on a ray system obeying Fermat's principle and boundary conditions based on Fresnel's formulas for reflection and refraction (so that both perfectly conducting and dielectric bodies can be handled). The scattered field is then determined essentially by the ratio of the size of a bundle of rays at the reflecting surface to the size of a bundle of rays at an observation point. Since this ratio can be made to depend on the principal radii of curvature of the reflecting surface at the reflection point, it is finite only for bodies with finite radii of curvature. In fact, if these radii of curvature are R_1 and R_2, then

$$\sigma_{\text{G.O.}} = \pi R_1 R_2. \tag{1.6}$$

For a sphere, in particular, $R_1 = R_2 = a$, so that

$$\sigma_{\text{G.O.}}(\text{sphere}) = \pi a^2. \tag{1.7}$$

Figure 1–1 shows how well the exact sphere RCS is approximated by equation 1.7; clearly, the approximation improves as a/λ increases. We shall discuss later the remaining curves on this figure.

The physical-optics method makes use of the geometric-optics field, as an approximation for the field on the surface of a scatterer, in the Stratton-Chu integral representation for the scattered field. In some cases the physical-optics result (which, unlike the geometric-optics approximation, depends on wavelength) is an improvement over geometric optics; for example, it permits a treatment of the finite flat plate, and the results are in very good agreement with experimental data. Figure 1–2 (from Kouyumjian [3]) gives an

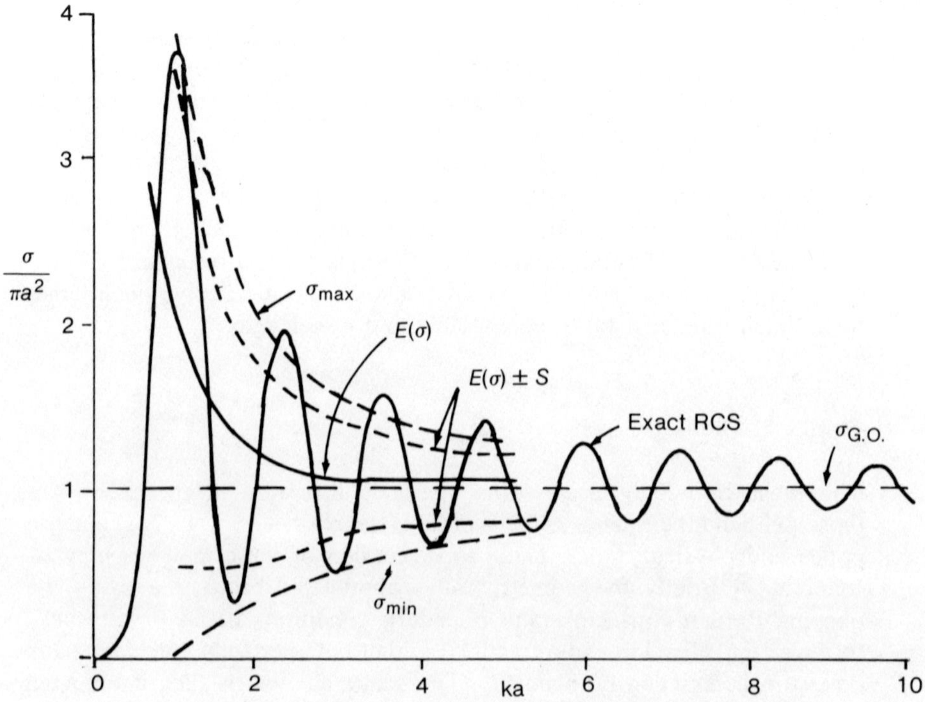

Note: G.O. = geometric optics calculated value; E = average value; S = deviation; RCS = radar cross-section.

Figure 1–1. Exact and Estimated RCS for a Sphere

illustration of the comparison. A comparison of both geometric-optics and physical-optics results with exact theory and experimental data is given in figure 1–3 for the prolate spheroid. Physical optics even gives good results for some bodies that violate the method's requirements of smoothness and large radii of curvature; for example, tip scattering from a cone and edge scattering from a wedge are handled nicely by physical optics (see [2], esp. pp. 94ff concerning ogives).

But the physical-optics method has the drawback that surface fields in the shadow region of a scattering body are assumed to be zero. For some scatterers, notably the finite cone and the cone-sphere, the RCS return may be essentially dominated by the effects of the shadow-region fields, at certain aspects and frequencies. Keller [4] and Fock [5] have developed approximation methods specifically designed to account for the surface currents that arise in the shadow region of a scaterer. The first is called *geometric theory of diffraction* (GTD); the second is usually called *creeping-wave theory*.

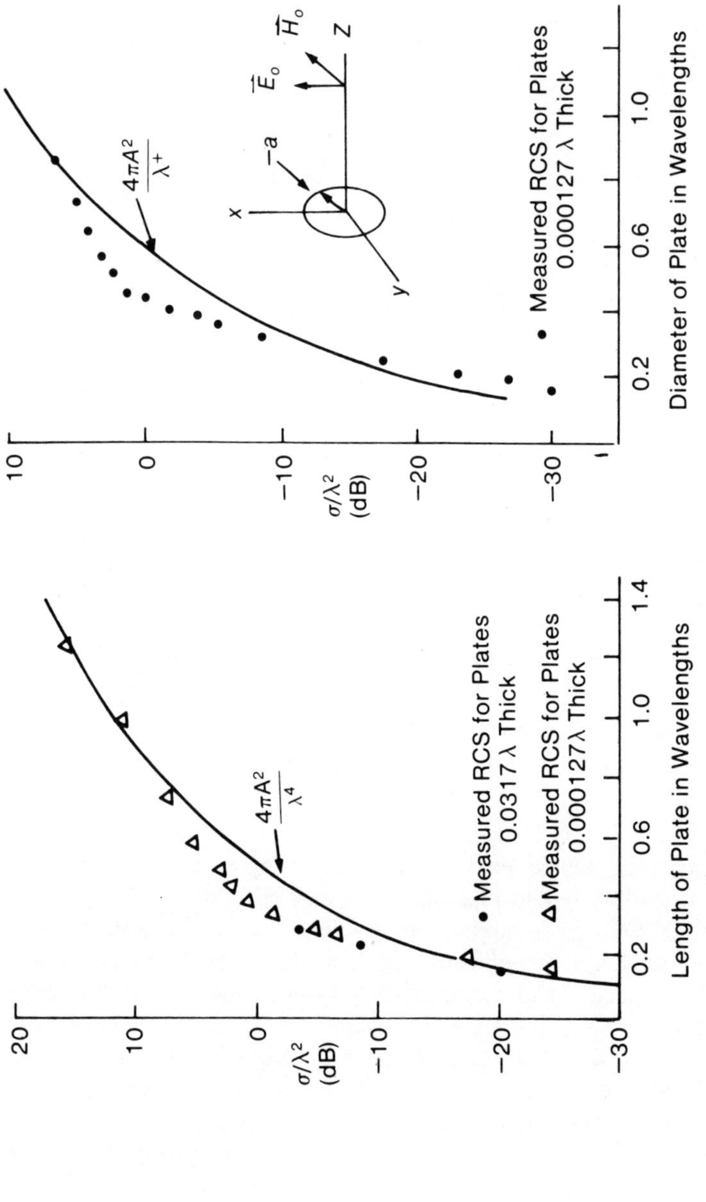

Figure 1–2. Echo Area of Thin Square (left) and Circular (right) Flat Plates at Broadside Aspect

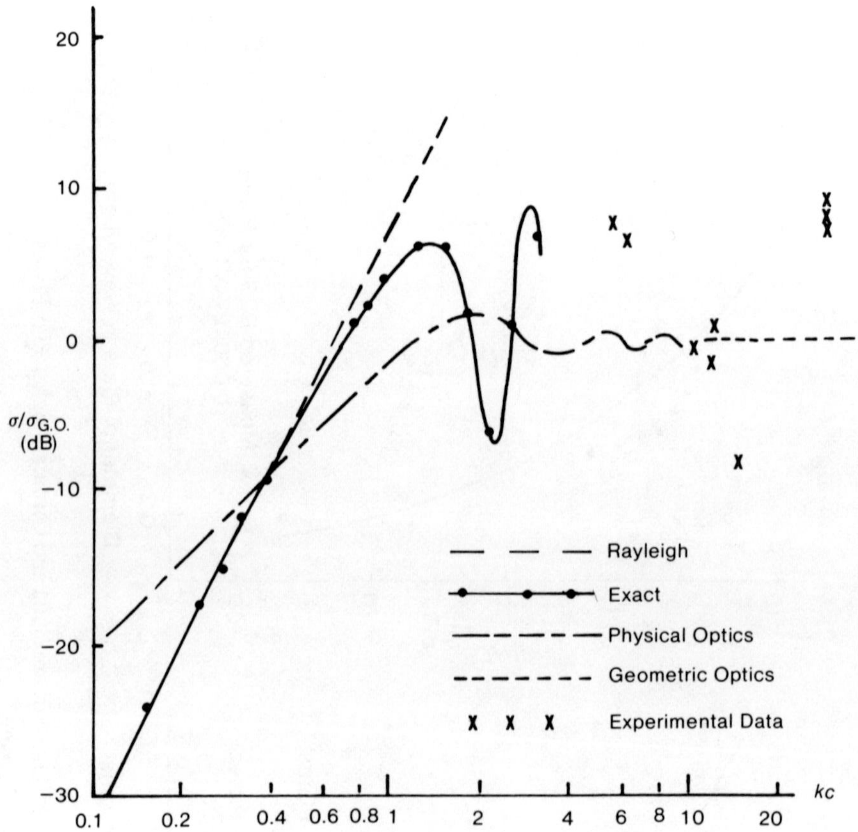

Note: $k = 2\pi/\lambda$; c = semimajor axis.

Figure 1–3. Nose-On RCS for a 10:1 Prolate Spheroid

Each has been developed to a fairly high degree of usefulness. We will discuss them briefly and illustrate what they will do.

The GTD can be applied to bodies that are smooth or that have discontinuities such as points or edges. To deal with the latter, for example, Keller introduced his so-called diffracted rays, which are produced when an incident ray strikes an edge. By using the optical principles of energy conservation and phase variation (as they are used in ordinary geometric optics), Keller was able to use *diffraction coefficients* to make a quantitative association between the field on a diffracted ray and the field on a corresponding incident ray. His diffracted rays are assumed to be straight lines that satisfy a *law of edge diffraction*: that is, a diffracted ray and corresponding incident ray make equal angles with the edge at the point of diffraction. These conditions on a diffracted ray are a result of Fermat's principle, which for the present purpose can be stated as: *a singly diffracted ray connecting two points is a curve whose length is stationary among all curves connecting these two points and having one point on the edge in question.* Thus it follows that one incident ray produces infinitely many diffracted rays, and that these diffracted rays lie on the surface of a cone having as axis the tangent to the edge at the point of diffraction. To find the diffraction coefficients, Keller appeals to an exact or asymptotic solution of a canonical problem—one whose essential scattering feature matches the one on the body under study. For example, to find the diffraction coefficient for an aperture, or a wedge, Keller appeals to Sommerfeld's exact treatment of diffraction of a plane wave by a screen in the form of a half-plane. In this way he is able to find the form of the diffraction coefficient that is applicable to both the case of vanishing field on the surface (Dirichlet problem) and the case of vanishing normal derivative of the field on the surface (Neumann problem).

Examples of what the GTD can and cannot do are given in figures 1–4 and 1–5. Figure 1–4 (from [6]) compares experimental data for a finite cone of 15° half-angle and 2.87λ base diameter with computations based on the GTD (as corrected by Bechtel [6]). Figure 1–5 shows that for the cone-sphere the agreement between experimental data and GTD results is perhaps not as acceptable as the agreement between experimental data and physical-optics results.

The creeping-wave theory introduced by Fock, (which first appeared in English around 1946 (see also [5]), is an attempt to deal with the effects of surface currents present in the shadow region of a body. The restrictions are similar to those of Kirchhoff theory. The obstacle must be convex and sufficiently smooth (in contrast to the points and edges admitted by GTD); and the wavelength must be very small in comparison with obstacle dimensions and radii of curvature of it surface. Fock carries out his analyses for monochromatic plane-wave excitation of the target. His fundamental result is the proof that the field has local character near the geometric shadow boundaries of the obstacle. That is, the field has an approximate value dependent only on the value of the external field at the point of investigation, the geometric characteristics of the surface element, and the electrical properties of the surface. Thus the current distribution near the shadow boundaries of the obstacle can be expressed through a universal function (identical for all bodies) whose argument is the ratio of distance from the geometric shadow boundary to the width of the half-shadow penumbra region. This being the case, Fock was able to determine the universal function by solving the canonical problem of plane-wave diffraction by a paraboloid of revolution. The theory is particularly adaptable to the cone-sphere, and figure 1–6 shows its application by Senior [7] to a 25° cone-

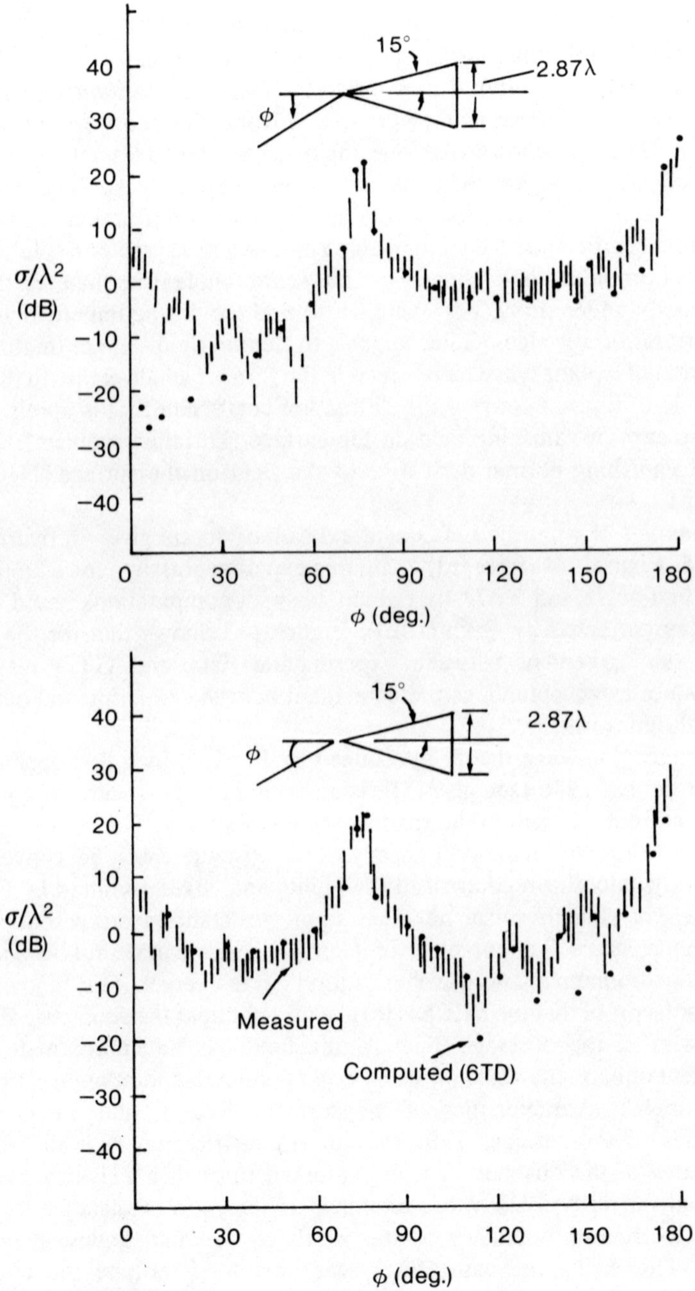

Note: Upper: vertical pol.; lower: horizontal pol. [6].

Figure 1–4. RCS for a Cone with Base Diameter 2.87λ ($ka = 9.02$)

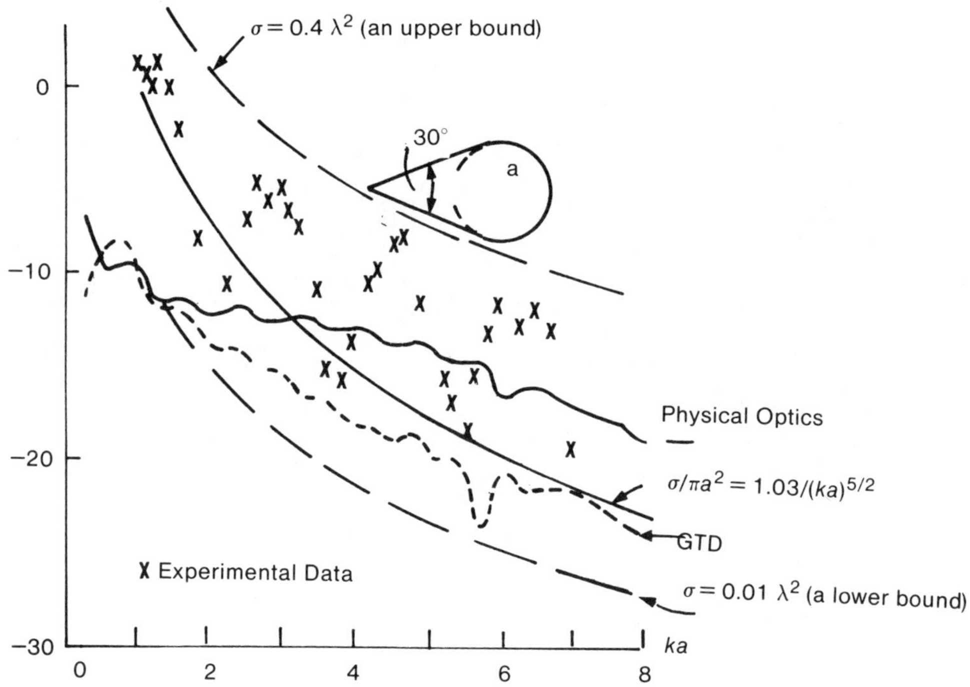

Figure 1–5. Nose-On RCS for a Cone-Sphere

sphere of base radius 4.519 cm in comparison with experimental data. (The cone-sphere is also a good shape from which to see the origin of the name *creeping-wave theory*. Fields that travel along a cone generator to the shadow boundary enter the shadow region where they "creep" around the spherical rear to the shadow boundary again and are launched back toward the observer).

The foregoing are the major approximate methods for finding the RCS of simple shapes. They can be combined, even for a single simple shape. Thus, for example, geometric optics and creeping-wave theory may be combined to give an RCS estimate for the sphere. Recently, Knott and Senior have been studying combinations of physical optics and GTD [8,9]. Indeed, even approximations of these approximations can be made and combined according to complex-body treatment outlined by equations 1.1–1.4. To illustrate this situation, the creeping-wave contribution for a sphere may be approximated by $1.03\ (ka)^{-5/2}$ and then combined with the geometric optics return πa^2 from its specular point. The average, rms spread, maximum, and minimum range as a function of ka are shown in figure 1–1; the estimate so obtained is quite acceptable.

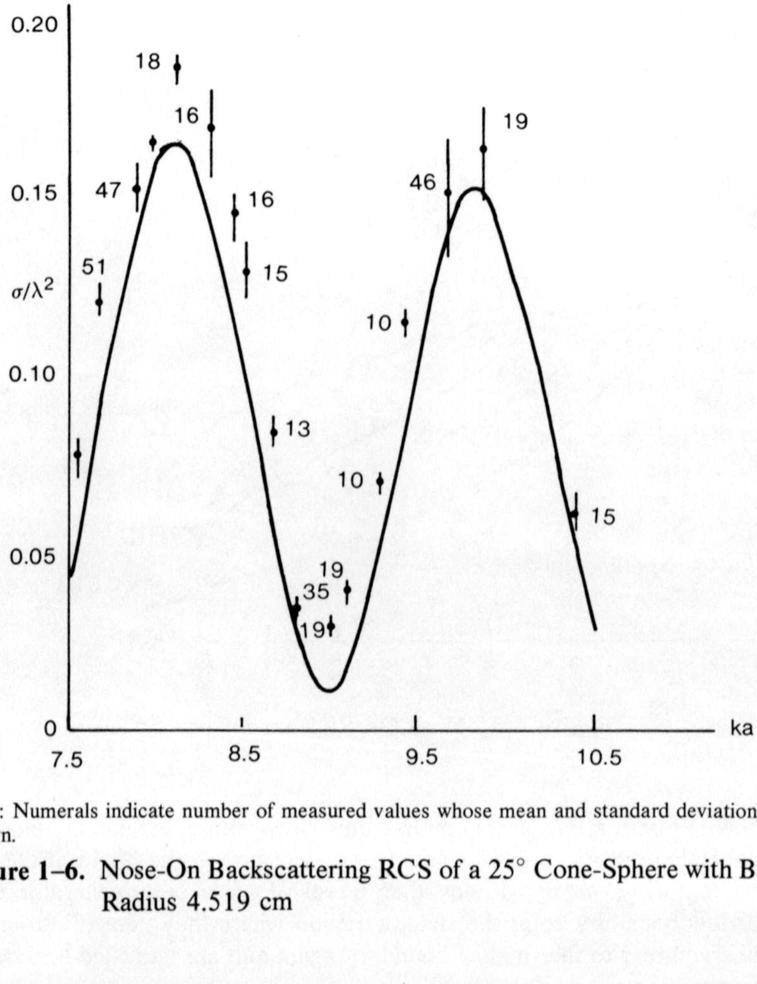

Note: Numerals indicate number of measured values whose mean and standard deviation are shown.

Figure 1–6. Nose-On Backscattering RCS of a 25° Cone-Sphere with Base Radius 4.519 cm

With these ideas in hand, it might be desirable to modify somewhat our definition of simple shape. Rather than have this phrase refer to a specific geometric shape, it is often convenient to have it mean the source of a particular type of RCS return. For example, we may wish to speak of the *creeping-wave component* when we mean the RCS return originating in the shadowed portion of a convex target, whose RCS contribution can be approximated by Fock's theory. In these circumstances a sphere would not be a simple shape but would be composed of two scattering components as indicated earlier: the specular point and the creeping wave from the shadowed region. A cone-sphere (viewed from the sharp end) would comprise three components: the cone tip, the cone-sphere join, and the shadowed (or creeping-wave) region.

We can now apply these methods to complex targets such as missiles and aircraft and can evaluate the resulting estimates by comparing them with experimental data. Consider first a fictitious missile whose size and shape are indicated by figure 1–7. Theoretical estimates of $E(\sigma)$ and σ_{max} have been made at 10.0 GHz. The complex scatterer was broken into six simple shapes: the inlet duct, the exhaust duct, the traveling-wave component (actually not a geometrical "shape"), the fuselage, the wings, and the stabilizers. The wings and stabilizers were broken into even simpler components by taking them to be collections of edges (modeled by thin wires). The fuselage too was broken into the simpler components of truncated ogive rear portion, cylindrical center section, and spheroidal nose. Figures 1–8 and 1–9 compare the average theoretical estimate of RCS with experimental data measured at 8.7 GHz. The discrepancy between this latter and the theoretical frequency of 10 GHz should not have unduly detrimental effect on the comparisons. Figure 1–8 shows a horizontal plane cut with the incident electric vector in the measurement plane; figure 1–9 shows a vertical cut, again with incident electric vector in the measurement plane. Differences between the theoretically estimated average value and experimentally measured envelope of peaks do no exceed 5 dB. Thus the agreement would improve if the experimental peak curve were lowered by several dB to allow for its being an envelope of maxima.

As an example of the application of the techniques being discussed to the estimation of RCS for an aircraft, we consider an exercise carried out some years ago for the Convair 990 commercial airliner [2]. It is not unlike the fighter, bomber, and reconnaissance aircraft we have studied over the past twenty years by these methods. Experimental measurements were made on a 1/25 scale model at 18 GHz, and the RCS estimate (of rms spread only) was made for the same model at the same frequency from a breakdown of the complex target into eighteen simple component scatterers. Dimensions of components were estimated from the model itself and not from drawings. Nevertheless, the agreement between theoretical estimate and experimental data is acceptable. The results [2] are reproduced here as figures 1–10 and 1–11.

Both of the foregoing examples have illustrated the random-phase method (equations 1.1–1.3). The relative-phase method (equation 1.5) is also appropriate when good dimensional information is available for phase calculation and when the largest of those dimensions is not much greater than the wavelength. One example of this approach for a missile shape similar to that of figure 1–7 has been given [2] and is reproduced here as figure 1–12. The agreement between theory and experiment speaks for itself. Many more examples for missiles, aircraft, reentry-vehicle shapes, satellites, and so forth can be found in [1] and [2].

In most cases where RCS is a highly oscillatory function of aspect,

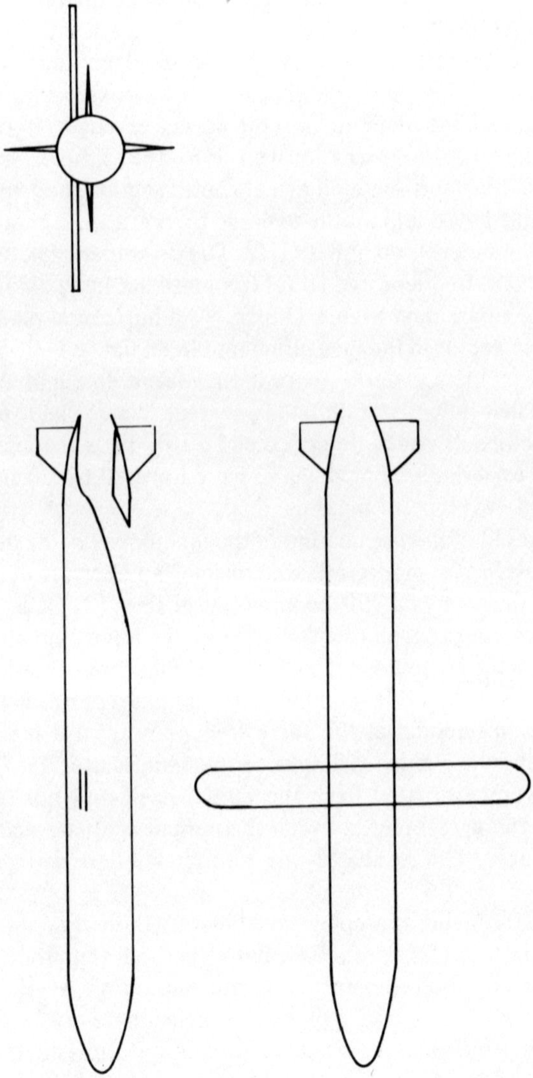

Note: 10:1 length-to-diameter ratio, say length 200" and diameter 20".

Figure 1–7. Configuration of a Hypothetical Missile

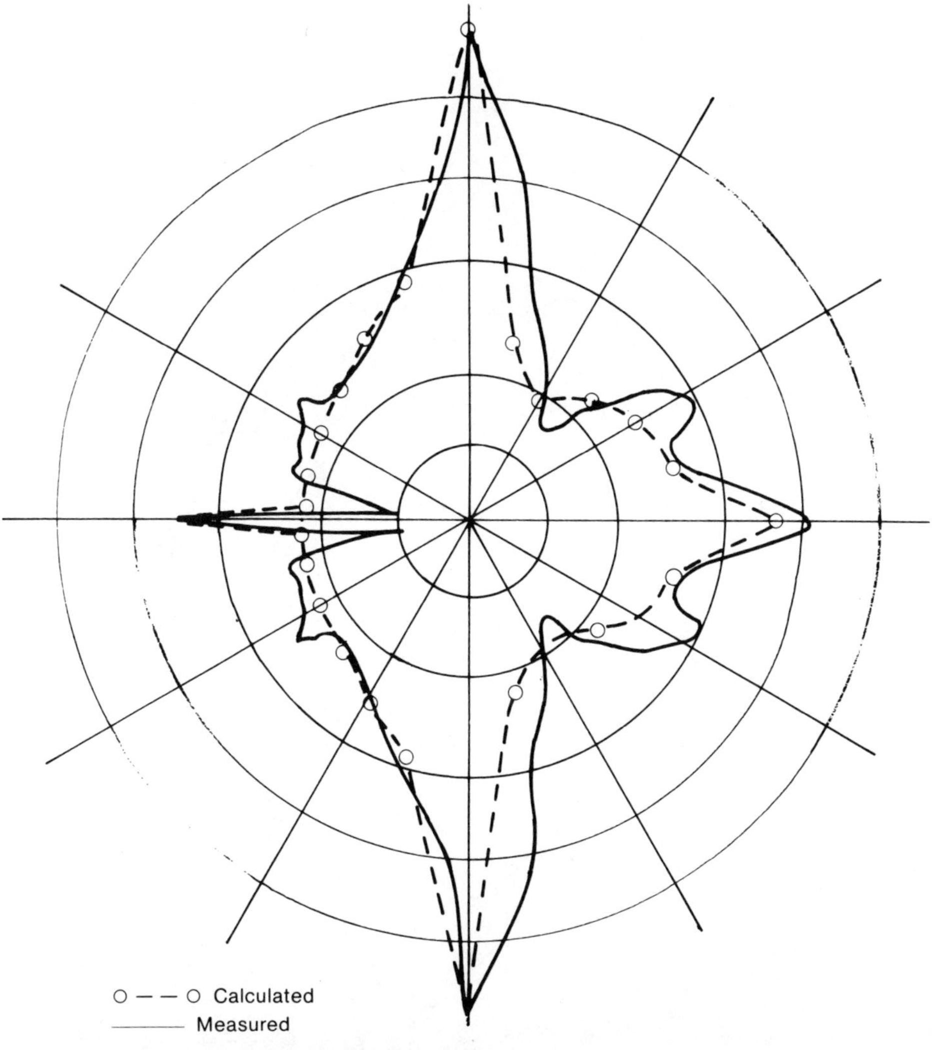

Note: calculated, 10 GHz; measured, 8.7 GHz

Figure 1–8. Calculated and Measured RCS in the Horizontal Plane with Polarization Vector in that Plane for the Hypothetical Missile

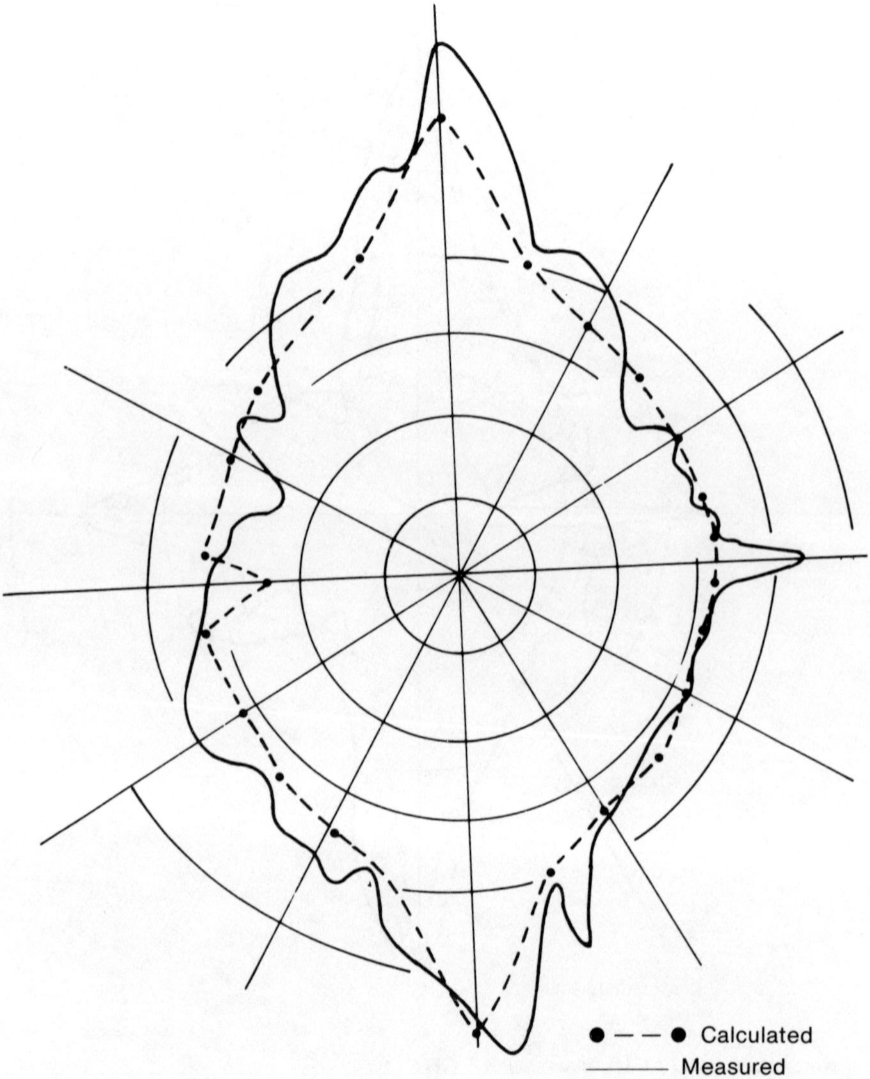

Note: Calculated, 10 GHz; measured, 8.7 GHz

Figure 1–9. Calculated and Measured RCS in the Vertical Plane with Polorization Vector in that Plane for the Hypothetical Missile

Figure 1–10. The Convair 990 (1/25 Scale)

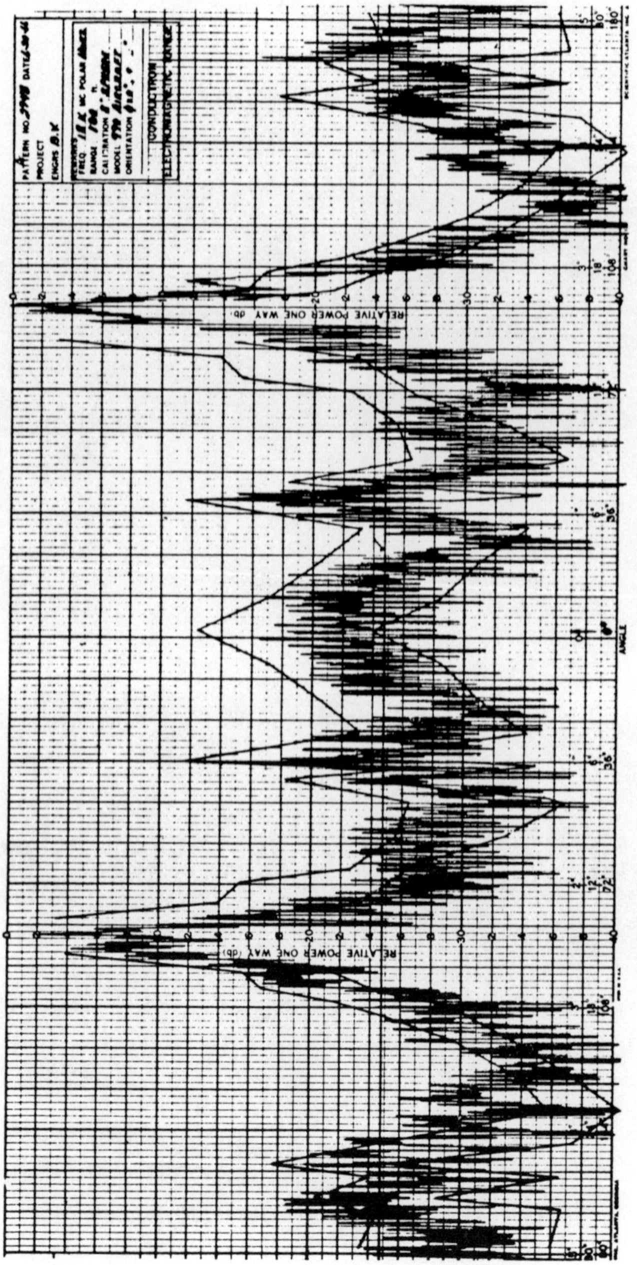

(A) hor. pol.

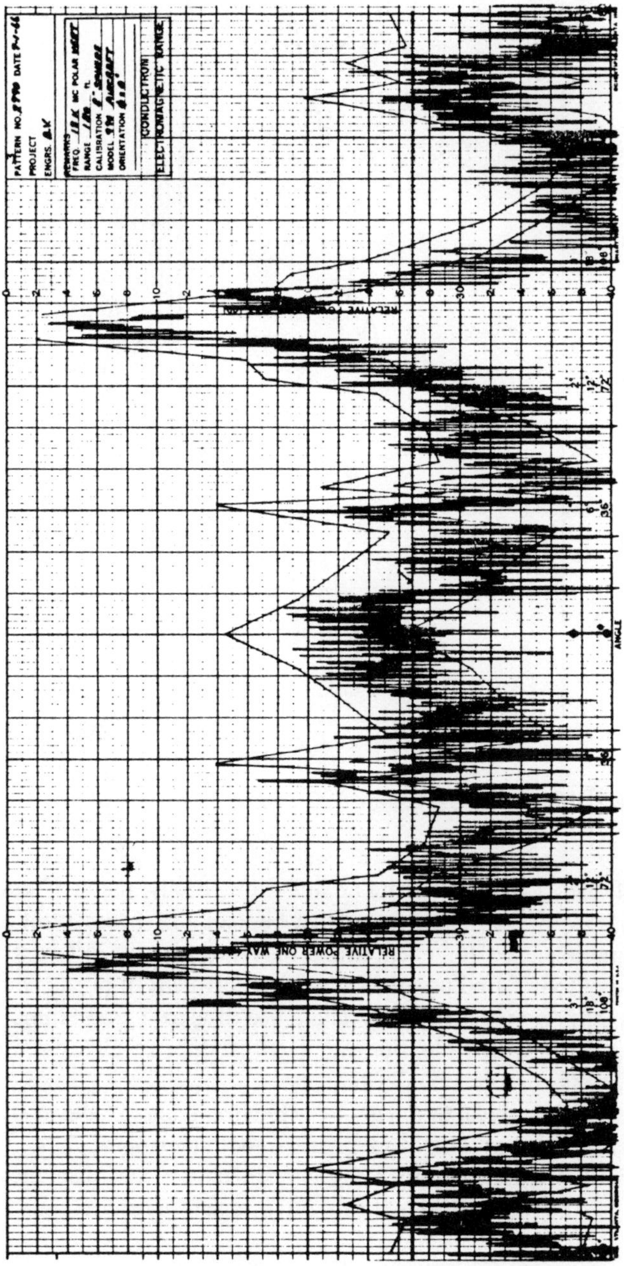

(B) ver. pol.

Figure 11–1. Calculated (rms Spread) and Measured RCS for a 1/25 Scale Convair 990 at 18 GHz

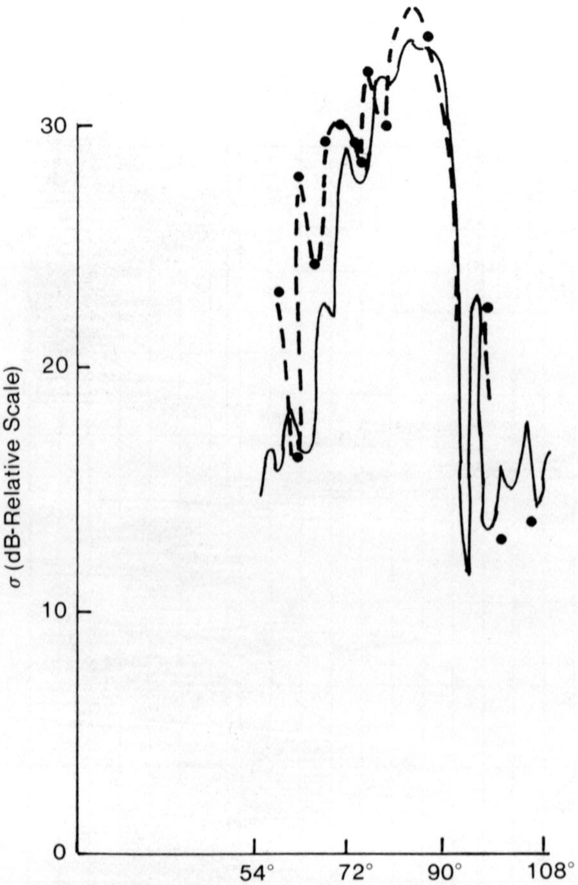

Figure 1–12. Near-Broadside RCS for a Reentry Shape

comparisons of theory and experiment are more easily made if each is smoothed in some way. But some care must be used to take account of the method of smoothing used for each, since they are usually different. The random-phase method of finding an average is equivalent to phase smoothing. In contrast, experimenters usually make use of aspect smoothing. Both methods are perfectly acceptable, but they may produce results that appear to be more divergent than is actually the case.

There are two examples that can be used to illustrate the contrasts between these aspect-smoothing and phase-smoothing approaches. These two, the flat plate and the cylinder, have been chosen because they are representative of the geometric shapes often used to model scattering components of missiles and aircraft. Moreover, although they are "simple" scatterers, the flat plate and the cylinder have all the features necessary to demonstrate the variance that can occur between aspect-smoothed and phase-smoothed results, provided they are sufficiently large compared with wavelength.

Consider first a square flat plate of edge length $a = 10\lambda$ located symmetrically in the *xy* plane with edges parallel to the coordinate axes. For the polar angle θ in the *xz* plane (that is, azimuth angle $\phi = 0°$) the RCS of the plate is well estimated by the method of physical optics to be

$$\sigma = \frac{4\pi a^4}{\lambda^2} \left| \frac{\sin(ka \sin \theta)}{ka\, sin\, \theta} \right|^2 \tag{1.8}$$

for $0° \leq \theta \leq 35°$. For example, from equation 1.8 we calculate that the first ten nulls occur for

$$\theta = 2.86°, 5.74°, 8.63°, 11.54°, 14.48°, 17.46°, 20.49°, 23.58°, 26.74°, 30°.$$

From an experimental measurement (figure 1–13), we observe that the nulls occur for

$$\theta = 3°, 6°, 9°, 12°, 15°, 18°, 21°, 24°, 27°, 30°$$

(as nearly as they can be read from the data).

However, in looking more closely at equation 1.8, we see that it can be interpreted as *two* scatterers (the edges), each having an RCS given by

$$\frac{a^2}{4\pi \sin^2\theta}$$

and being combined by the phase-relation sin (ka sin θ). Therefore, to get a phase-averaged estimate for the plate RCS, we simply appeal to the reasoning given earlier (see equations 1.1–1.3) for finding the average RCS for two or more components (remembering that this technique requires the assumption that all phase angles are equally likely). Therefore, our phase-averaged, or smoothed, estimate for the plate RCS becomes

$$E(\sigma) = \frac{a^2}{4\pi \sin^2\theta} + \frac{a^2}{4\pi \sin^2\theta} = \frac{a^2}{2\pi \sin^2\theta}. \tag{1.9}$$

Notice that $E(\sigma)$ will run 3 dB below the envelope of peak RCS values for the plate. The graph for $E(\sigma)$ is superposed on the experimental curve in figure

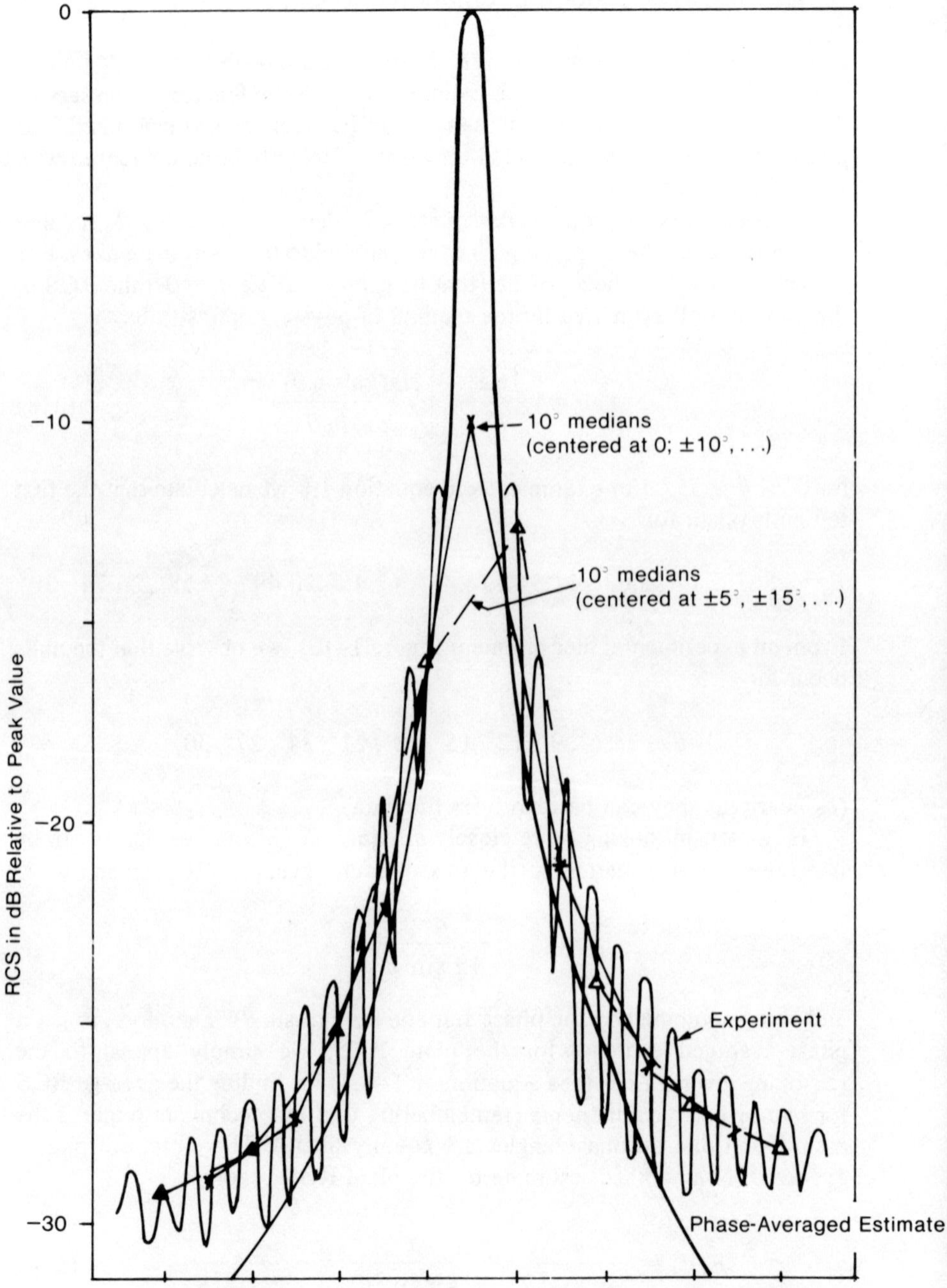

Figure 1–13. Comparison of Measured RCS with Phase-Averaged Estimate and Medians over 10° Intervals for a Square Flat Plate 10λ on a Side

1–13. Of course, $E(\sigma)$ "blows up" for $\theta = 0°$; so we avoid this catastrophe by simply fairing its curve into the value $4\pi a^4/\lambda^2$, which is the limit for σ in equation 1.8 as $\theta \to 0°$. Perhaps to be consistent with the fact that $E(\sigma)$ lies 3 dB below the peak values of σ, we should fair $E(\sigma)$ into $2\pi a^4/\lambda^2$. As we shall show, this variation would make very little difference in the comparison of $E(\sigma)$ with the aspect-smoothed RCS σ_{10m}.

The form of aspect smoothing to be considered here is the computation of medians in a 10° interval of data, which can be of any type but is usually, and in particular is here taken to be, expermental measurement data. Thus, in a given fixed 10° aspect interval, we find that RCS value σ that is exceeded by other RCS values in 50 percent of the interval. Such σ are usually plotted at the midpoint of the interval and conected by straight-line segments. For the experimental data given in figure 1–13, we have computed medians over 10° intervals centered at 0°, ±10°, ±20°, and ±30° and have plotted the medians σ_{10m} at the midpoints of these intervals.

We see then in figure 1–13 a 10-dB discrepancy between the aspect-smoothed and phase-smoothed approximations for the experimentally measured values in the vicinity of $\theta = 0°$. Yet in all other respects the approximations are in reasonable agreement. Such difficulties, it appears, will always occur when a large and very narrow peak occurs in the data, in particular when the peak is narrow with respect to the interval over which medians are taken. The aspect-smoothed approximation will fall considerably below such a peak, whereas the phase-smoothed approximation, which will usually involve a fairing into the actual peak value, will usually hit the experimental peak almost exactly. In fact, if the scattering component has an effective scattering aperture of about 10λ, it will have about a 5° beamwidth at the 10-dB level. Thus, if medians are taken over 10° intervals, σ_{10m} will always be about 10 dB lower than $E(\sigma)$ (providing the return of said scattering component is essentially isolated from all other returns). On the other hand, if one removes the peak situations, agreement between results from the two smoothig methods should be generally quite good.

To illustrate further the problems introduced into aspect-smoothing results by peaks, we observe that discrepancies can be made to occur between separate aspect-smoothed results if the aspect intervals over which medians are taken are shifted. For example, for the aforementioned plate we have also computed medians over 10° intervals centered at ±5°, ±15°, ±25°, and ±35°, again plotting the medians at midpoints of the intervals. Figure 1–13 shows the discrepancies that arise between this aspect-smoothed result and the one just presented. As before, it is the height and narrowness of the peak that cause the trouble.

As a second example apropos of the scatterers exemplifying missiles and aircraft, we consider briefly a cylinder of diameter $2a = 3.37\lambda$ and length $L = 10.94\lambda$. Again we can see (figure 1–14) that the effective scattering

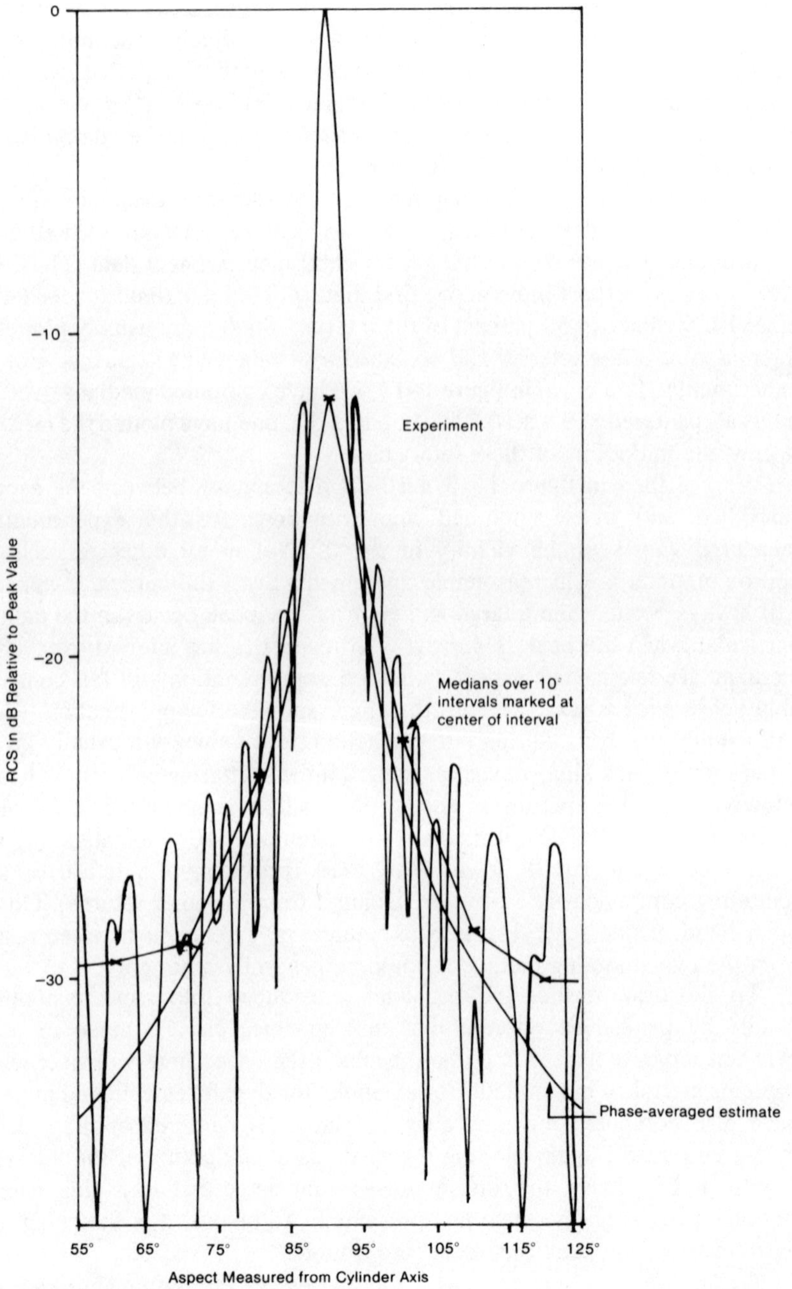

Figure 1–14. RCS of a Cylinder of Length 10.94λ and Diameter 3.37λ

aperture, in this case length, of about 10 wavelengths causes about a 10-dB discrepancy between aspect-smoothed and phase-smoothed estimates of the cylinder RCS.

In this example also we have a "simple" scatterer that, for off-normal directions, can be viewed as two scatterers (the cylinder ends), each with RCS

$$\frac{a\lambda \sin \theta}{8\pi(\cos \theta)^2},$$

combining with a phase sin (kL cos θ), θ being measured from the cylinder axis. As we saw in the case of the plate, the foregoing expression does not exist at $\theta = \pi/2$ and so must be faired into the normal incidence value kaL^2. Thus we again have the situation where the phase-smoothed peak value approximates the experimentally measured peak value very closely, whereas the peak value of the aspect-smoothed approximation will fall about 10 dB below the experimentally measured peak value. For other aspects up to 30° off normal, the results of the two methods will be in acceptable agreement.

Thus it is clear that some care and awareness of potential difficulties must be exercised when comparisons are to be made between theoretical estimates of RCS (usually phase-smoothed) and experimental estimates of RCS (usually aspect-smoothed) at aspects where RCS varies by a large amount with change in aspect; at other aspects this type of comparison is quite valid.

Rayleigh and Resonance Methods

Let us turn now to those situations for which wavelength is much larger than a major dimension of the target. We refer to this region, from the frequency viewpoint, as the *Rayleigh* or *low-frequency region.* It is possible to expand the scattered field amplitude in positive powers of the wavenumber k ($= 2\pi/$wavelength) with coefficients that are functions of aspect, polarization, and the body. Little is known about the convergence of this series, although it is thought to converge if the body is finite and k small enough (for example, for a sphere of radius a, the radius of convergence is defined by the value of k such that $ka = 1$). The first term of the series can be ascribed to the electric and magnetic dipole contributions to the scattered field, and yields Rayleigh scattering (hence the name of the region) for which RCS varies as k^4. This kind of dependence has been used by Siegel [10] to suggest an effective approximation for Rayleigh RCS. Viewing the scattering body as a singularity in the field, so that scattering is essentially indpendent of body shape, he uses dimensional analysis to observe that the coefficient of the k^4 term must be proportional to the square of the volume of the body. To account only for body elongation, he also used a shape factor chosen to fit the RCS for an ellipsoid. The result is

$$\sigma = k^4 V^2 \left(\frac{4}{\pi}\right)\left(1 + \frac{1}{\pi y} e^{-y}\right)^2 \qquad (1.10)$$

where y is the length-to-width ratio of the body, which is valid for wavelength much greater than the major body dimension. The use of the so-called *shape-factor* $[1 + (1/\pi y)e^{-y}]^2$ can be modified to produce an aspect-dependent formulation of [10]. A comparison of experimental data with estimates obtained from [10] is made in figure 1–15 for a nose-on view of five finite cones of varying half-cone angle. It should be observed that agreement between theoretical estimate and experimental measurement as a function of base-circumference-to-wavelength ratio (between 0.1 and 10) improves as the cones become "fatter" (that is, as the half-cone angle increases). When the circumfrence-to-wavelength ratio is unity or less, it is impossible to distinguish between RCS levels for nose-on and base-on views.

The chief difficulty with Rayleigh region approximations of this type [10] is to set their upper frequency limit of usefulness. In fact, a similar statement can be made concerning the lower frequency limit of usefulness of optics methods. All of which is a backhanded way of introducing the so-called resonance region, which is generally rather ill defined and may differ from body to body. It is often taken to be centered at that frequency for which a major body-dimension-to-wavelength ratio is unity. For example, in the case of the sphere when the circumference-to-wavelength ratio is unity, we are at the peak of the first oscillation of the exact solution (see figure 1–16). Then, if we move to the left and right one-half unit, we find the Rayleigh approximation to be quite good in the first case and the optics approximation to be fairly good in the second (within 3 dB). Thus we might take as a working definition of the resonance region that interval where the Rayleigh and optics approximations begin to diverge from each other. Indeed, we can often press into the resonance region from the Rayleigh and optics ends to obtain RCS estimates where good approximation methods do not exist.

For example, consider the hypothetical missile shape given by figure 1–17 at 20 MHz. The length and wing span would then be approximately 0.4 and 0.2 wavelengths, respectively. Then, at nose-on and tail-on aspects, upper and lower RCS bounds could be gotten from the Rayleigh and optics approximations, respectively, for the body. At broadside the body RCS could be estimated as that due to a "thin wire." When the results are faired together, the range of values would be those shown in figure 1–17 for incidence direction in the plane of the wings, and for (electric) polarization vector in $(\hat{P}_2)$, and perpendicular to $(\hat{P}_1)$, the plane of constant roll angle. If the same routine is followed for the wings, the dotted ranges shown in figure

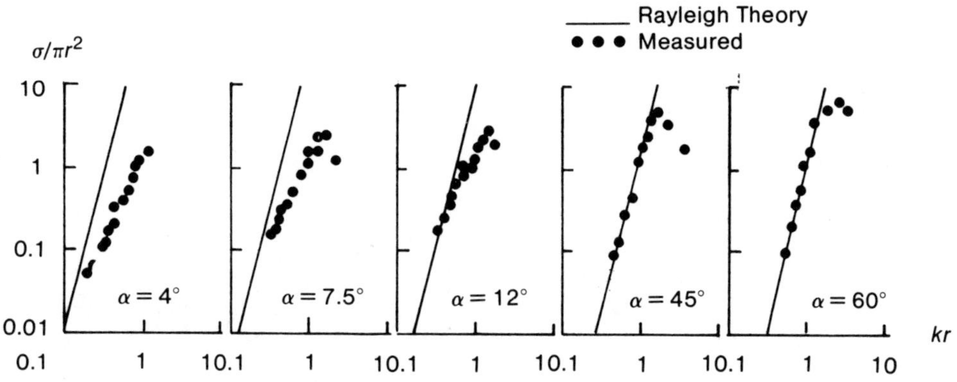

Note: $\alpha = 1/2$ cone angle; $2r =$ base diameter.

Figure 1–15. Nose-On RCS for Finite Cones in the Rayleigh Region

1–17 are obtained. Allowances must thus be made for the two types of RCS return to combine in some appropriate phase, or it will usually suffice to give upper and lower bounds by permitting them to combine in and out of phase, respectively.

Experience has shown many times that the same kind of approach also

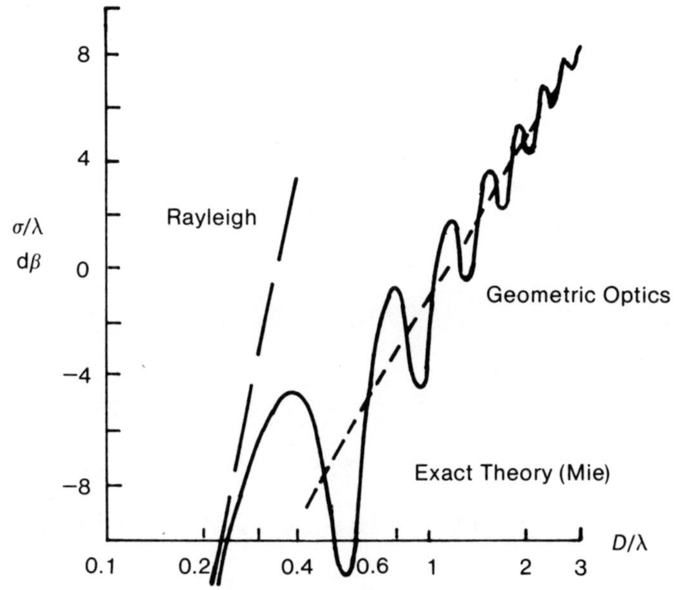

Figure 1–16. RCS for a Sphere of Diameter D

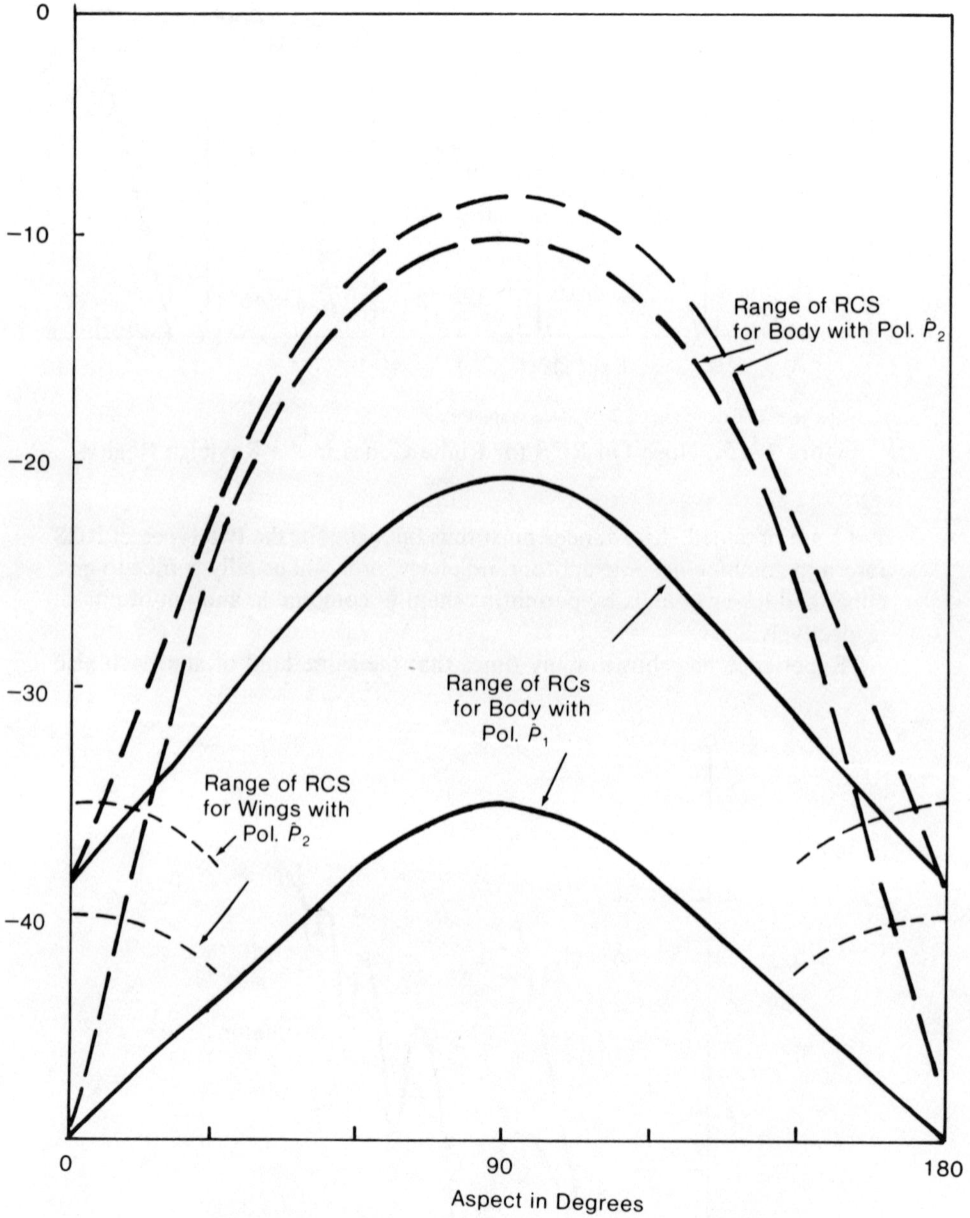

Figure 1–17. RCS for Hypothetical Missile at 20 MHz

works remarkably well for aircraft. Thus we see that although there are almost no approximation methods directly applicable in the resonance region, good estimates of RCS for this region can be obtained by a careful use of Rayleigh and optics methods; such estimates are always much aided by deductions from experimental data, where such are available. Moreover, it is still possible to use the individual scattering-component approach to the complex-body problem.

Summary of State of RCS Methods

It is possible to estimate the RCS of a complex scatterer by: (1) decomposing it into a collection of simple scatterers, where *simple* may indicate a type of scattering (such as creeping- or traveling-wave return) rather than a simple geometric shape; (2) computing the RCS for each member of the collection; and (3) combining the items from step (2) deterministically with suitable phases to get a relative-phase estimate (see equation 1.5), or statistically with an assumption of independent random phases to get a random-phase estimate (equations 1.1–1.4). These methods are valid for the optics region and into the resonance region (with care); Rayleigh-region estimates will not usually require any decomposition of a complex target (which will actually not appear complex to a Rayleigh wavelength). The theoretical RCS results for complex shapes such as aircraft and missiles will almost always agree with experimentally measured (static) RCS data to within 3–5 dB.

Bistatic RCS

Bistatic-RCS studies constitute only a small percentage of the total attention, both theoretical and experimental, that has been given to radar-reflectivity problems since World War II. A likely reason for this is that there exist results, again both theoretical and experimental, that indicate that many bistatic problems of interest can be reformulated as equivalent monostatic problems. The qualification on such equivalence is that the bistatic angle β between directions of illumination and observation be limited to an appropriate region, which is often governed by some major body-dimension-to-wavelength ratio but, in general, should not exceed 135°. The region between this limiting value, β_0, and $\beta = 180°$ might be called the near-forward scattering region wherein the bistatic RCS is usually considerably higher than it is for the domain $0 \leq \beta \leq \beta_0$. Exceptions to this statement may of course occur at certain aspects, particularly for complex targets where certain scattering components may be shadowed by others.

The theoretical equivalence between bistatic and monostatic scattering can be stated in terms of the so-called *bistatic theorem*:

> In the limit of vanishing wavelength, the bistatic RCS for illumination direction $\hat{k}$ and observation direction $\hat{n}$, $\hat{k} \neq \hat{n}$, is equal to the monostatic RCS for the illumination-observation direction $\hat{k} + \hat{n}$ for bodies that are sufficiently smooth.

The theorem is dependent on the assumptions made for physical optics—that is, that the current generated on the scatterer is to be approximated by twice the tangential component of the incident magnetic field on the illuminated portion of the surface and by zero on the shadowed portion.

This result has the usual drawback of a physical-optics approximation. Because the transition of surface current from illuminated to shadowed regions of the scatterer is not appropriately modeled, it is necessary to mitigate this error by restricting the use of the theorem to those cases for which the dominant scattering dimension is sufficiently large compared with the wavelength, and the scattering body itself is sufficiently smooth. In practice, however, it is possible to relax the former restriction considerably. For example, as we shall see when we examine some of the experimental results that follow, a major scattering-dimension-to-wavelength ratio as low as 20:1 is quite acceptable. A proof of the bistatic theorem can be found in [13].

Keller's geometric theory of diffraction (GTD), which is reviewed elsewhere in this discussion, is applicable to the larger bistatic-angle domain since scattered fields computed from GTD can be traced along any diffracted ray. No such effort has been carried out to my knowledge. However, the University of Michigan Radiation Laboratory has made studies [14, 15] of certain two-dimensional shapes (such as wedges and ogival cylinders) for bistatic angles up to 60° (for special angles of incidence); in this work particular attention was given to nonspecular scattering (which becomes important when specular scattering can be reduced) and to the reduction of bistatic RCS by surface modification (such as surface loading or impedance variation with RAM).

A theoretical result of the sort discussed earlier can be put on a more solid footing if it can be verified by experimental measurements of real targets of interest, such as aircraft and missiles. Such a course is available here. Crispin and Siegel [2] give bistatic-measurement results for an aircraft (bomber size) in both median and maxima format at a frequency of 250 MHz (horizontal polarization), and for a missile in the form of maxima and minima at a frequency of 200 MHz (horizontal polarization).

We discuss the aircraft bistatic data first. The bistatic angles considered for the aircraft are $\beta = 0°$, 30°, 60°, and 135°; the results are presented as a function of θ, which denotes the angle between the nose-on direction and the "equivalent monostatic direction" given by the vector $\hat{k} + \hat{n}$. The case $\beta = 0°$ is itself obviously the monostatic situation. Therefore, the result of the bistatic theorem should be verified if the RCS curves for all the indicated β values are essentially identical. Clearly this situation is indeed true for both the 10° median and 10° maximum values, provided it is agreed that a 10-dB range defines "reasonable agreement." If the curve (figure 1–18) for $\beta = 135°$ is omitted, the agreement among the RCS curves for the remaining three β values improves to a 5-dB range. It was essentially for this reason that 135° was chosen as an upper limit of validity for the introductory remarks on bistatic and monostatic equivalences.

The situation for the missile data (figure 1–19) is not quite as satisfactory as it is for the aircraft, partly because the β domain is increased ($\beta = 0°$, 40°, 90°, and 150°), and partly because the scattering phenomenа governing the RCS of a (long, thin) missile may be dominated by mechanisms (such as traveling waves; see also [14] and [15]) that are not well represented by the specular type of model provided by physical optics. In addition, the minima of a set of measured data do not furnish good criteria for evaluation or comparison since they are so sensitive to relative-phase changes among scattering components. Nevertheless, they do give here a pronounced indication of the deterioration of the bistatic theorem for $\beta = 150°$, and a very clear indication of the increase of near-forward RCS over monostatic RCS for most of the θ domain. Once again, if the data for this targets β value were omitted, the general agreement of the remaining data would be much improved.

The bistatic RCS for a body in the Rayleigh region can also be obtained theoretically by an extension of the monostatic results [2]. Bistatic data, measured in the Rayleigh region, are available for the sphere and finite circular cone (for a nose-on equivalent monostatic vector), and these results confirm theoretical estimates.

In summary, although bistatic results of both a theoretical and an experimental nature are minimal in comparison with monostatic work, the experimental data that do exist strengthen and confirm theoretical estimates based on the bistatic theorem for a bistatic angle domain $0 \leq \beta_0 \leq 135°$.

RCS Reduction

It sometimes happens that the initial design of a missile or an aircraft has RCS characteristics that are undesirable, necessitating the initiation of an RCS-control effort. Theoretically, one might wish to have a larger return, but usually the goal is to lower the RCS.

Fortunately, RCS-reduction priorities can usually be established that

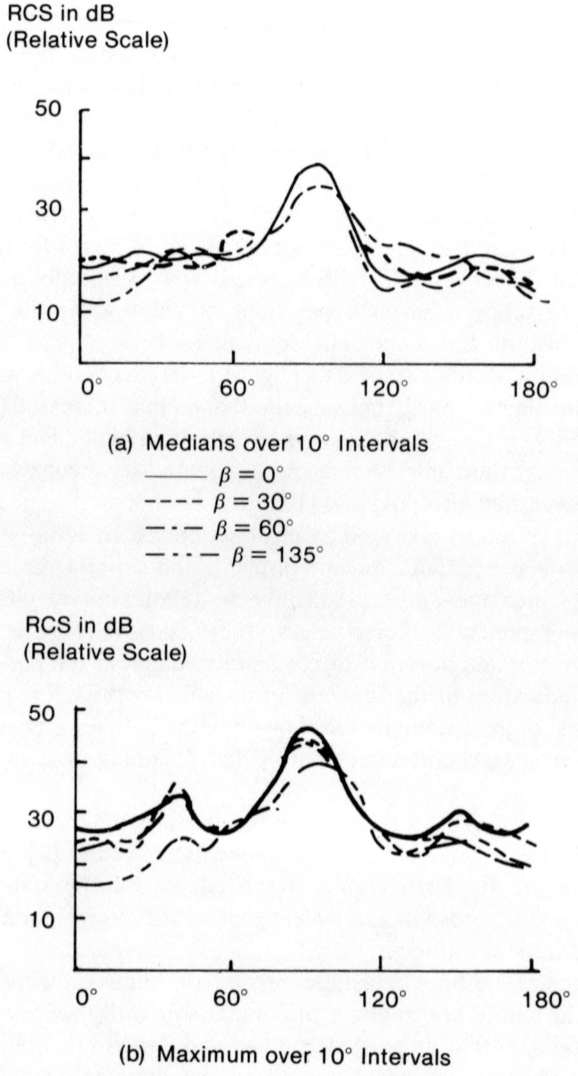

Figure 1–18. Bistatic RCS Data versus θ Measured at 250 MHz (Hor. Pol.) for a Large Jet Aircraft

make an RCS-reduction problem both feasible and practical. That is, usually some aspects are more important than others and some frequencies more crucial than others. For such situations one can often obtain a decided improvement in RCS characteristics through an RCS-control program.

There are two basic methods for RCS reduction:

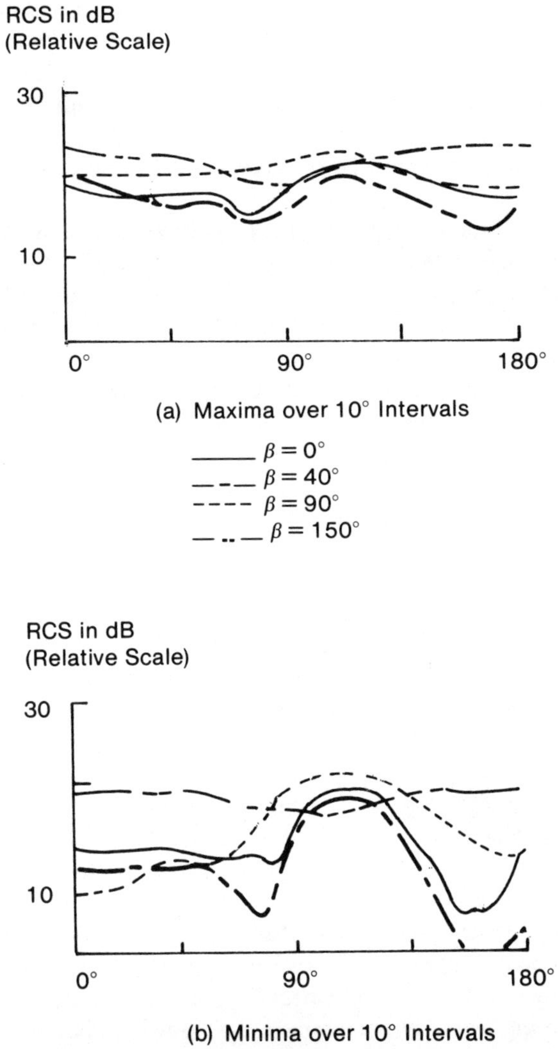

Figure 1–19. Bistatic RCS Data versus θ Measured at 200 MHz (Hor. Pol.) for a Missile Shape

1. geometric alteration of scattering configuration so as to redirect reflected energy from an aspect region of concern to one of little or no concern;
2. alteration of constituent parameters of scattering surfaces so as to make them essentially transparent (for example, fiberglass wings) or absorptive (for example, radar-absorbing material [RAM] coatings).

These methods are passive and are applied to passive scatterers. Two additional categories of RCS reduction must be added to deal with

3. active interference with the surface currents on a scatterer to induce RCS reduction;
4. RCS reduction for active (electromagnetically) targets.

Item 3 encompasses an area describable as impedance loading, where early studies have really been confined to passive methods. Thus, point 3 is a variation and combination of points 1 and 2. Item 4 deals with antennas and sensors. These areas are discussed briefly (particularly so for the last) in this sections.

It is true that each target must be considered as an individual situation, but some general comments can be made about the application of methods 1 and 2 to aircraft and missiles.

RCS Reduction by Geometric Modification

When RCS reduction is desired, it is obtainable only over a limited aspect region and then at a cost of increased RCS in some other aspect region. Moreover, if success is achieved by gaining considerable reduction of RCS for a given component scatterer, then care must be exercised to check whether other component scatterers, previously considered insignificant contributors, are now producing a significant RCS. As the frequency becomes low enough that body length or wing span is of the order of the wavelength, configuration changes become impossible. Thus there is a lower frequency limitation on this technique imposed by the body size; but the more the frequency increases, the more useful the method becomes.

Together with the frequency, the principal radii of curvature of the scatterer dominate the nature of the RCS pattern, that is, its amplitude and lobe structure as a function of aspect. There are essentially five cases into which scattering can be categorized by allowing the principal radii of curvature to range from zero to infinity:

1. If both radii are zero, the body will have a point, and the RCS will be proportional to λ^2 for near point-on aspects.
2. If one radius is zero and the other is finite or infinite, the body will have an edge, and the RCS will vary like λ for near edge-on aspects.
3. If both radii are finite, the body will be smooth, and the RCS behaves like λ^0 for all aspects.
4. If one radius is infinite and the other finite, the scatterer is cylindrical, and the RCS behaves like λ^{-1} normal to the cylinder axis.

5. If both radii are infinite, the surface is flat, and the RCS behaves like λ^{-2} for aspects normal to the flat surface.

Cases 4 and 5 are characterized by large peaks whose widths are governed by cylinder length or flat-surface extent; in fact, for off-normal aspects case 5 behaves like case 4. Case 3 produces an isotropic pattern (as long as the radii do not vary) whose amplitude is proportional to radii size. Usually, cases 1 and 2 produce the smallest RCS at higher frequencies.

These five cases offer the ideas for RCS modification. It must be emphasized again, however, that they do not reduce the RCS but merely trade it from one aspect region to another. Thus a rounded missile nose could be made pointed. Then, at near nose-on aspects, the RCS would be greatly lowered; but at aspects far enough off nose-on, the RCS will increase and finally exceed the value it had for the rounded nose.

Wing edges could be curved instead of straight. The RCS will then be (nearly) uniform over all edge-on aspects, in contrast to the RCS for the straight edge, which will have a higher peak at edge-on aspects but will be much lower at off-edge-on aspects.

Mesh-screen covers are often suggested for ducts. Suitable orientation will cause them to give at the normal aspect a RCS much higher than the duct RCS but, at all other aspects, an RCS much lower than the duct RCS.

A long missile fuselage might be changed from a cylindrical to a rectangular parallelopiped if very high RCS could be tolerated at four isolated aspects (normal to the paralleopiped faces) in exchange for very low RCS elsewhere in a plane normal to the missile axis.

RCS Reduction by Material Modification

For absorber use in reducing RCS, there are also aspect and frequency limitations. First, the off-normal absorptivity performance of a RAM does not usually match its normal incidence performance, falling off in direct proportion to an increase in angle of incidence off normal. Second, it is sometimes the bistatic performance of an absorber that should be of optimum quality, such as, for example, is desirable for its use in an inlet duct. Here again it is generally the rule that as the size of the bistatic angle increases from 0° (normal incidence), the absorptivity performance of the RAM worsens. Here, as with geometric modification, difficulties attendant on RAM use increase as radar wavelength increases. It is generally true for all RAM that the lower the frequency for which the RAM is designed, the thicker—and usually heavier—the RAM must be. An exception to the increase in weight with an increase in thickness for RAM can be seen in the "Salisbury screen" absorber (a very thin conducting layer with a resistivity of 377 ohms placed one-half wavelength from a perfectly conducting surface), which in fact can be made of material no heavier than a sheet of blotter paper located an appropriate distance from a metal surface. Obviously, however, such RAM could have no applicability for any body that must move in the atmosphere (although it might on a bulkhead behind a canopy or radome). Frequency bandwidth over which a material can peform efficiently as an absorber is generally rather narrow if its thickness and weight are to be kept within reasonable bounds. Bandwidth may, however, be increase at an appropriate cost, such as by layering materials designed to perform in different frequency bands, or by using ceramic materials of increased complexity (the incorporation of magnetic properties), which is usually accompanied by increased weight and thickness.

Because of the weight penalty and frequency selectivity of RAM, the geometric-modification technique should be considered first, particularly if there is some aspect priority involved in the RCS-control problem.

A detailed discussion of RAM behavior has been given by Bowman [2]. But a brief discussion of the mechanism by which RAM may be made to reduce RCS is worthwhile because, of course, the type of absorptivity mechanism chosen directly affects such vital parameters as frequency coverage and weight penalty.

To act as an absorber a material must first admit impinging energy; hence it should have a low reflection coefficient or a normalized front-face impedance as near unity as possible. Once inside the material, the advancing energy must be attenuated or absorbed; thus its wave number, within the material, times the distance it travels in the material must have a reasonably large imaginary part. Part of these latter requirements can be met by a material with a large index of refraction modulus, a condition that implies that the energy within a plane slab (that is, a single layer) of such material will travel in a direction nearly normal to the slab [11]; indeed, it is this condition that justifies a careful study of the impedance for normal incidence only. Reasonable confidence in an equivalence between specular and nonspecular monostatic performance (up to angles of 40° off normal) of radar-absoring material (RAM) has been established; absorption characteristics determined by flat-plate free-space tests (or, equivalently by slotted-line tests) give reasonable estimates for the performance of RAM placed on a cone sphere where, for near-nose-on aspects, the radar cross-section is determined by surface-current effects.[1]

For the energy to travel a reasonable distance in the material, the slab thickness should be many wavelengths (as measured in the material). Two extreme cases are immediately evident: the thick- and the thin-layer cases. The former occurs when the slab thickness is much greater than the material wavelength, or skin depth. Here the admitted energy will find an ample distance in which to be attenuated and will be admitted into the material (that is, the material will have a low reflection coefficient), provided only that the material permeability and permittivity are approximately equal [12].

The thin case occurs when the slab thickness is small compared with the free-space wavelength. Here the reflection coefficient is small if the imaginary part μ'' of permeability is large compared with the real part μ'. The larger the ratio μ''/μ', the larger the slab thickness becomes compared with the material wavelength; thus even though the layer is thin, it provides an increasing effective attenuation distance with increasing μ''/μ'. Between these two extremes the reflection coefficient and attenuation are complicated functions of the complex material-constituent variables, permeability and permittivity. In this interval special "resonant" cases arise when the layer thickness is an integer multiple of quarter or half wavelengths (again, as measured in the medium). With ferrite absorbers this is a condition often manifest in narrow-band, deep-null performance of the reflection coefficient.[2] Resonant situations are not suitable for large-bandwidth absorbers but are excellent for reasonably narrow-band performance.

Often the thin and resonant conditions can be suitably combined to produce fairly wideband performance. Thus a ferrite absorber can be designed to give a resonant performance at a high frequency, say in the X-band region. Its quarterwave thickness is then sufficiently thin to make it perform as a thin absorber in, perhaps, the *S*-band region. An example of such wideband performance is given by Emerson-Cuming "Eccosorb FGM," which has for a thickness of 3.17 mm (2.5 lb./ft^2) an average reflectivity of -12 dB over the 2–12-GHz band.[3]

There is little that can be done at 3 to 30 MHz (HF) frequencies, since a RAM designed for that band would in all likelihood be too thick and heavy for practical use.

We conclude this discussion by considering a few simple modifications, both geometric and material, that could be encorporated in the hypothetical missile model introduced in figure 1–7. The intent is not to offer the last word in missile design but only to show that considerable reduction in RCS is realizable through a modest number of alterations. The modified version, where RCS will be estimated at 17.5 GHz, is shown in figure 1–20. The nose shape has been changed from prolate spheroidal to ogival. The wings and tail shapes have been modified so that leading and trailing edges ar not parallel. It is also assumed that the wing and tail material has been changed from metal to a fiberglass having a low dielectric constant. The inlet duct has been elongated with the entryway sloped so as to permit a minimum of incident energy to strike directly on the compressor face. It is also assumed that RAM with effective absorptivity of 15 dB has been properly located in the duct. (Each individual duct requires its own analysis for a suitable location of RAM; usually, careful location of RAM will make its effective absorptivity some 5–10 dB better than its nominal absorptivity.) Thus it is assumed that

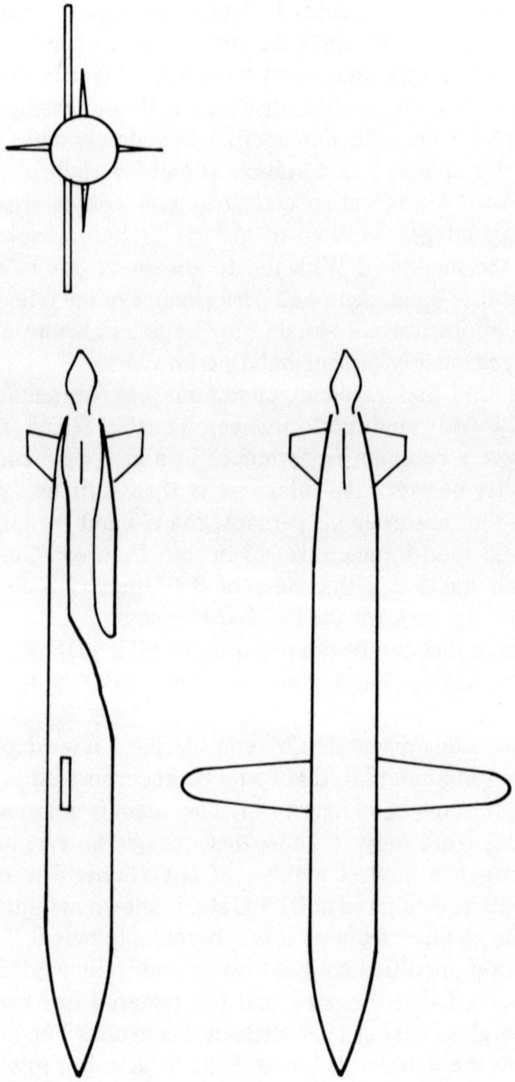

Figure 1–20. Configuration of a Hypothetical Missile Modified to Reduce RCS

the duct-throat shaping and RAM modification together afford a 20-dB reduction of duct RCS. Finally, an ogival insert has been located in the exhaust duct, leaving room for the escape of exhaust gases but preventing a direct view of the exhaust duct.

No account has been taken of traveling-wave (TW) contributions, since they are insignificant at the high frequency being considered here. If a much lower frequency were to be considered, then a TW wave lobe at near-nose-aspects would have to be dealt with. To reduce this TW contribution would require either a RAM coat on the rear portion of the fuselage or some other treatment to interrupt or absorb the traveling waves. In the next section we consider briefly the idea of impedance loading in this connection.

The estimated reduction of RCS realizable at 17.5 GHz from the geometric and material modifications is shown in figure 1–21 for a vertical plane cut under the missile. We see that, other than at near-broadside for which no modification was considered, the geometric and material changes can be quite effective in RCS reduction.

RCS Reduction by Impedance Loading

A variation of RCS reduction by geometric and RAM modifications that seeks to alter surface currents on a scattering body is adaptive RCS control by impedance loading. By an adaptive system we mean one that is programmed to adapt itself to some well-defined sequence of RCS outputs as a function of frequency, aspect, polarization, and so on; or one that is self-adaptive in the sense of responding in particular ways to particular frequency, aspect, or polarization stimuli. We can confine ourselves to the first, since the second will depend almost entirely on a successful approach to the first.

The remarks in this section are still in the nature of comments on the state of the art of RCS-modification ideas. Here, however, the ideas are somewhat more speculative than those discussed for geometric shaping and RAM application, and certainly do not enjoy such a vast literature of favorable test results.

Programming the response of a scatterer so that it adapts itself to some preset values of such parameters as time, range, altitude, and so on requires the control of scattered fields by adaptive impedance loading. To achieve a broad range of control would undoubtedly require the distribution of impedance over an array of loads. Such techniques might even offer advantages in RCS control over what is obtainable from geometric modifications or surface coatings. We will examine the state of the art briefly and look at one or two directions from which this area might be reasonably be entered.

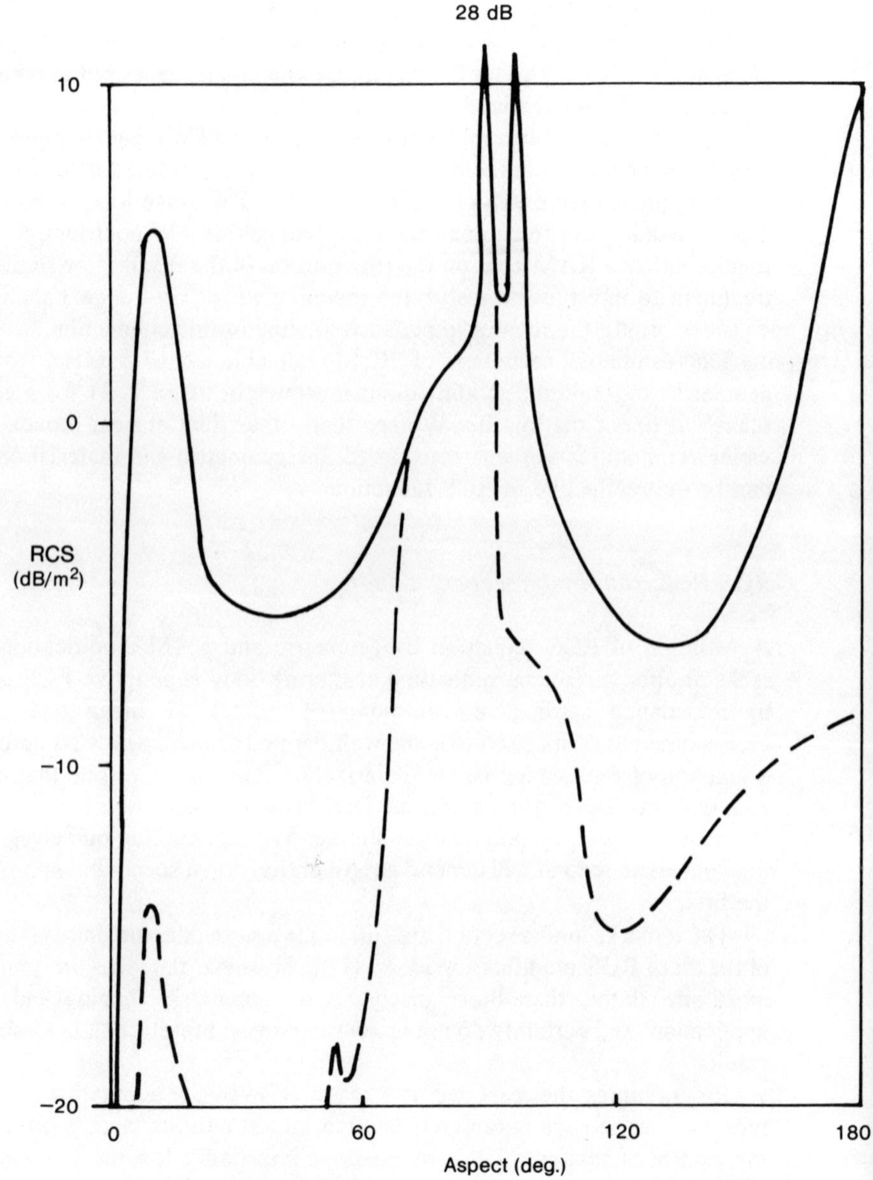

Figure 1–21. Estimated RCS for Hypothetical Missile

The literature contains only a very meager set of facts on the subject of impedance loading for the purpose of controlling the RCS of scattering bodies. Only geometries of a very special nature have been considered to date, and the loading as applied to these geometries is also very special. For example, in 1965 Liepa and Senior [16] considered the problem of the sphere loaded by a nonuniformly excited circular solt located in a plane perpendicular to the direction of incidence. Noting that a nonnegative real part of loading admittance corresponds to passive loading, these authors point out that zero RCS is not uniformly achievable over all ka (a being the sphere radius and k being the wave number) with a passive loading, or with an active loading, because of the oscillatory behavior of the real and imaginary parts of loading admittance as a function of ka.

Chen [18] considered in 1966 the reduction of RCS in the end-fire direction of a long thin body. Chen's treatment of the long thin body constitutes a clever physical interpretation of the mathematics involved. He shows that a load can be introduced in such a way as to cancel the waves traveling forward from the rear of the body, so that only forward-traveling waves are supported between the front of the body and its load. Between the load and the rear of the body, "the backward wave cannot be cancelled because a standing wave is needed to satisfy the boundary condition at the rear end of the body.

There is essentially no information extant on the subject of broadbandedness and polarization effects. Liepa and Senior [16] remark that "any reactively loaded device such as the one considered [by them] will be severely limited in its bandwidths unless more sophisticated techniques, nonlinear or even active, are employed in the synthesis of loading." Chen finds, [18] particularly for the long thin body, that the placement of the load is very heavily dependent on frequency and that this is why its effect is very narrow band. Schindler, Mack, and Blacksmith [17] have given a good review of material extant on the RCS loading problem up to 1965. One of the approaches considered in the following discussion is based on some of the conclusions of these authors.

Schindler [17] has pointed out that the scattered field from a loaded object may be thought of as the superposition of two fields, one representing the known field of the unloaded body (the unloaded scattered field) and the other representing the radiation effects of the loading structure (the loaded scattered field). One possible approach that could be developed is as follows. Approximate methods for calculating the RCS of perfectly conducting bodies are well established and very successful. It might be possible to develop a similar theory to approximate the RCS of impedance-loaded objects. Then, following the basic statement of Schindler et al. [17], we would superpose fields proportional to these cross-sections with phases appropriate to the frequency and aspect involved.

As an illustration of approximation methods that could provide engineering estimates of RCS, consider the following treatment of the loaded sphere, which then can be compared with the work of Liepa and Senior [16]. Let the direction of incident energy be along the polar axis of a sphere with a loaded slot located on the equator of the sphere. Maximum and minimum relative backscattering cross-sections for passive loading have been calculated by Liepa and Senior and are given in figure 4 of their paper [16]. We propose to model the contribution from the passive equatorial slot by the first- and second-order contributions from the base of a finite cone for nose-on backscattering as determined by Keller [4]. The second-order or diffraction term of Keller's result contains a phase factor with argument of the form $2ka$, where a denotes the cone base radius, which arises from energy traversing the cone base and being diffracted toward the nose-on direction. To adapt this result so as to model the equatorial slot on the sphere, we would replace the phase argument $2ka$ by an argument $2kd$, where d would represent the slot depth.

The Keller terms would now provide an estimate of the impedance-loading contribution from the equatorial slot to the RCS of the sphere. This result may now be combined both in and out of phase with the specular return from the sphere to obtain maximum and minimum estimates, respectively, for the loaded-sphere RCS. The maximum estimate is illustrated in figure 1–22 and compared there with the Liepa and Senior results. Clearly, the estimate is not perfect. However, we feel that, as a first cut, it is sufficient to illustrate the efficacy of the idea and the model. Questions remain concerning methods of interpreting the outputs of these RCS models in terms of impedance loads. That is, phasing of RCS return may be determined so as to achieve a desired RCS control as a function of frequency and aspect; but such phasing must then be realizable in terms of arrays of impedance loads. These matters will be treated briefly.

The work from the literature referred to here has also made use of approximations. For example, both Chen [18] and Liepa and Senior [16] have relied on the fact that the loading slot is small with respect to the wavelength in at least one dimension. Rather than lift this restriction so as to achieve a distributed impedance loading, it seems more natural to retain it, and the results obtained using it, and to turn toward a study of a collection or array of such loading configurations over the surface. Such a method of implementing a programmed adaptive system of RCS control would also be practical in the sense of minimizing the surface area over which such loading would have to be distributed.

Another initial direction in which to proceed, particularly for missile shapes, might be based on a *body-perturbation method.* This approach would attempt to extend Chen's results by "fattening" one end of the long thin body in an attempt to find that point at which the Chen theory begins to

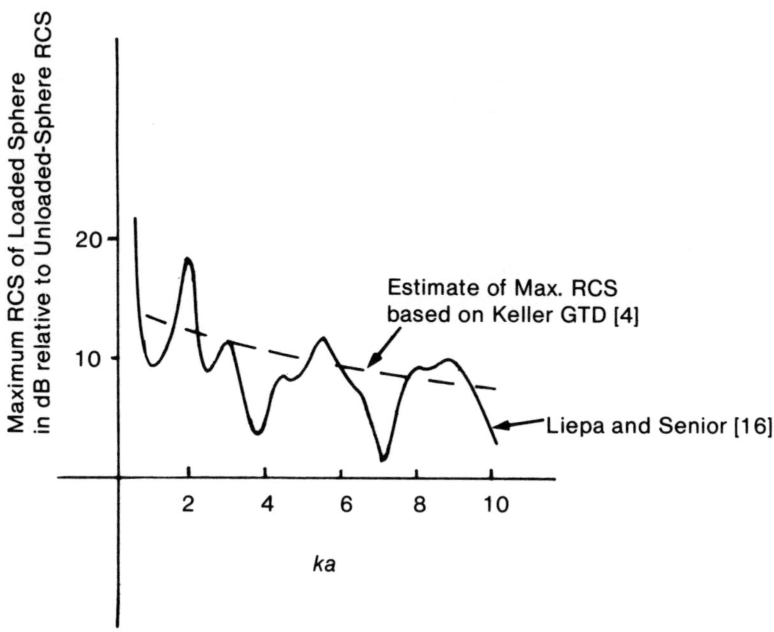

Figure 1–22. Comparison of Exact Solution with an Approximate Solution for a Passively Loaded Sphere

break down. This would have the advantage of maintaining and perhaps extending Chen's physical interpretation of the waves traveling on the body. In any case it would be valuable to keep track, if possible, of the bistatic situation involved. Clearly the monostatic situation is of chief interest, and we wish to develop a loading that would produce a desired null or peak in some direction; but if we could do so, it would be valuable to be able to direct the energy in some unimportant direction.

One can intuitively conclude that a network of lumped impedance loadings might extend the frequency coverage of the concepts described earlier. If so, a certain degree of polarization insensitivity would surely result. The impedance network can be viewed as a discrete approximation to a continuous-surface impedance loading and hence represents a compromise between the continuous and point impedance-loading techniques.

The scattering that results from a network of lumped impedance loadings that are separated on the scatterer by distances less than the longest wavelength of interest requires a different analytical approach. One interesting possibility is to apply the theories of surface-wave launching and propagation. In this procedure the impedance network is viewed as a periodic distributed network with a propagation constant determined by common techniques. For example, assume that the single load of the dipole treated by Chen is replaced by a series of loads over some portion of its length. The analysis of the endfire echo can, in principle, be obtained just as before if the propagation constant is employed for the loaded dipole section. This loading would also modify the current distribution, but this would not present any insurmountable difficulties.

More general networks, with impedances that are distributed in a manner analogous to the distribution of elements of frequency-independent and broadband antennas, could conceivably be treated in a like manner and lead to improved RCS-control systems.

RCS Reduction for Antenna and Sensor Configurations

Future attention might profitably be directed to the RCS return from antennas and sensors. It would appear to be intrinsically impossible to reduce the RCS of antennas and sensors at their operating frequencies. At other frequencies, however, RCS reductions can be obtained.

The methods to be used should, no doubt, be governed by the particular type of antenna, or sensor, that must be camouflaged, and by the system in which it is operational. There are two distinct types of approach. First, the reflector and/or feed configurations can be contructed in special ways. Second, the radome can be specially constructed, using either a suitable grid of wires embedded in it or a coating that passes certain frequencies and reflects others, giving in either case a band-pass radome.

A considerable literature already exists on these subjects, and an appropriate departure point might be a review of such methods within the constraints imposed by the other missile-systems requirements.

Conclusions

The subject of RCS estimation has by no means been exhaustively treated in the foregoing sections. Much effort and many volumes have been given over to the subject. More detailed discussion, as well as a much more complete bibliography, may be found in [1] and [2]. Treatments of the scattering problem from the viewpoint of exact solutions for simple shapes may be found in [19] and [20]. Related areas that would require special treatment, such as scattering from ionized configurations, have been omitted entirely.

We have, however, in the time and space available, attempted to give an idea of the working tools available for the task of RCS estimation for air vehicles, what kinds of problems they can handle, and how good such estimates are in comparison with measured data. The result is that the RCS of almost any complex scatterer can be estimated well enough to agree with experimentally measured data to within 3–5 dB, where *complex* includes material constituency as well as shape.

What areas require more attention? This is a difficult question; in my opinion, however, estimation formulas for RCS returns from certain simple, or component, contributors—such as a cavity and a traveling wave—could enjoy a firmer footing, although the scattering mechanisms of each is well understood in principle. The inlet duct, for example, may present such a bewildering variety of shapes and material variations as to defy any ray-tracing schemes, which are likely to be inappropriate for RAM-coated walls in any case. The traveling wave depends not only on the parameters of the surface over which it propagates, but also on the reflection coefficient of the "termination" from which it reflects. The determination of a usable reflection coefficient depends largely on good intuitive use of experimental data. Both these areas could profit greatly from careful and extensive experimental measurement programs. The data could be used, at least, to generate improved empirical estimation formulas, at most to improve understanding of the "simple" scatterers. It is intended, of course, that such areas of study should include the material variations imposed by RAM use.

Bistatic-RCS estimation methods for complex targets at large (say, greater than 135°) bistatic angles are essentially nonexistent. This area very much needs work, both theoretical and experimental. Even for smaller bistatic angles, very little is known about the existence and behavior of simple scatters—for example, traveling waves. The same dearth of knowledge exists for the effects of material constituency in bistatic scattering. There is evidence, for instance, [19] that absorbers become enhancers of RCS for bistatic angles in excess of 135°.

Finally, there is the area of antenna and sensor camouflage. Although methods exist for RCS reduction at frequencies other than the operating frequency, little is known about what RCS reductions, if any, may be possible at the operating frequency. It would be helpful to have experimental measurements that would check out existent schemes for RCS reduction at all fequencies of interest with the antenna or sensor in actual operation at its operating frequency.

In closing, we should like to call attention to a very helpful document for those interested in RCS target signatures: an extensive bibliography [21] of radar and infrared camouflage reports; it begins with work done in the 1950s.

Notes

1. See also, for example, Emerson and Cuming, Inc., "Preliminary Technical Bulletin 8-2-18, Eccosorb SF," 20 January 1969: "Typically an X-band version has a thickness of 0.20″, weighs 2 pounds/sq. ft., and will

exhibit normal incidence reflectivity of −25 dB at the resonant frequency." Furthermore, it "will at that frequency maintain reflectivity of at leas −20 dB out to 30° incidence and at least −16 dB out to 45°."

2. The only practical resonant ferrite absorbers for which this observation has been made have been at most of quarter-wavelength thickness.

3. See Emerson and Cuming, Inc., "Preliminary Technical Bulletin 8-2-23, Eccosorb FGM," dated 8 January 1969.

References

1. Crispin, J.W., Jr., and Maffett, A.L. "Radar Cross Section Estimation for Complex Shapes." *Proc. IEEE* 53 (August 1965): 972–982; also, "Radar Cross Section Estimation for Simple Shapes." *Proc. IEEE* 53 (August 1965): 833–847.
2. Crispin, J.W., Jr., and Siegel, K.M., eds. *Methods of Radar Cross Section Analysis.* New York: Academic Press, 1968.
3. Kouyoumjian, R.G., "Calculation of the Echo Area of Several Scatterers of Simple Geometry by the Variational Method." *Proc. Symp. Microwave Optics.* Eaton Electronics Research Laboratory, McGill University, Montreal, 1953.
4. Keller, J.B., "Geometrical Theory of Diffraction." *J. Opt. Soc. Amer.* 52 (February 1962): 116–130; "Backscattering from a Finite Cone-Comparison of Theory and Experiment." *IRE Trans. on Ant. and Prop.* AP-9 (July 1961): 411–412; "Backscattering from a Finite Cone." *IRE Trans. on Ant. and Prop.* AP-8 (March 1960): 175–182.
5. Fock, V.A. "The Field of a Plane Wave Near the Surface of a Conducting Body." *J. Phys. USSR* 10 (1946) :399–409. This work is one of a collection of Fock papers edited by N.A. Logan and P. Blacksmith, Jr., ASTIA Document No. AD 117276 (June 1957): "Diffraction, Refraction, and Reflection of Radio Waves," Thirteen Papers by V.A. Fock.
6. Bechtel, M.E., "Application of Geometric Diffraction Theory to Scattering from Cones and Disks." *Proc. IEEE* 53 (August 1965): 877–881.
7. Senior, T.B.A. "The Back Scattering Cross Section of a Cone-Sphere." *IEEE Trans. on Ant. and Prop.* AP-13 (1965): 271.
8. Knott, E.F., and Senior, T.B.A. *Non-Specular Radar Cross Section Study.* University of Michigan Radiation Laboratory, Technical Report AFAL-TR-73-2, February 1973.

9. Knott, E.F., Liepa, V.V.; and Senior, T.B.A. *Non-Specular Radar Cross Section Study*. University of Michigan Radiation Laboratory, Technical Report AFAL-TR-73-70, April 1973.
10. Siegel, K.M. "Far Field Scattering from Bodies of Revolution." *Appl. Sci., Res. B* 7 (1958): 293–328.
11. Uberall, H. "Radar Scattering from Coated Perfect Conductors: Application to the Semi-Infinite Cone and Use of Exact Eikonal." *Radio Science,* 69D (June 1964).
12. Weston, V.H. "Theory of Absorbers in Scattering." *IRE Trans. on Ant. and Prop.* AP-11 (September 1963).
13. Siegel, K.M.; Alperin, H.A.; Bonkowski, R.R.; Crispin, J.W.; Maffett, A.L.; Schensted, C.E.; and Schensted, I.V. "Bistatic Radar Cross Sections of Surfaces of Revolution." *J.A.P.* 26 (1955): 297.
14. Knott, E.F., and Senior, T.B.A. "Non-Specular RCS Study." AFAL-TR-73-2, February 1973.
15. Knott, E.F.; Liepa, V.V.; and Senior, T.B.A. "Non-Specular RCS Study." AFAL-TR-73-70, April 1973.
16. Liepa, V.V., and Senior, T.B.A. "Modification to the Scattering Behavior of a Sphere by Reactive Loading." *Proc. IEEE* 53 (August 1965): 1004–1011.
17. Shindler, J.K., and Mack, R.B., eds. "Special Reports no. 6, the Modification of Electromagnetic Scatterings Cross Sections in the Resonant Region—A Symposium Record." vols. I and II, AFCRL-64-727 (1 and 2), Microwave Physics Laboratory Project 5635, 4 June 1963.
18. Chen, Y.M. "Diffraction by a Smooth Transparent Object." *Journal of Mathematical Physics* 5 (1964): 820.
19. Ruck, G.T. (ed.) "Radar Cross Section Handbook," vols. 1, and 2. New York: Plenum Press, 1970.
20. Crispin, J.W., Jr., and Siegel, K.M. (eds.) "Methods of Radar Cross Section Analysis." Academic Press, 1968.
21. Corriher, H.A., Jr., and Pyron, B.O. "A Bibliography of Articles on Radar Reflectivity and Related Subjects. 1957–1964" *Proc. IEEE* 53, no. 8 (August 1965), pp. 1025–1064.

Appendix 1A: General Comments on Control and Reduction of Radar Cross-Section

It sometimes happens that the initial design of a missile or an aircraft has RCS characteristics that are undesirable, necessitating the initiation of an RCS-control effort. Theoretically, one might wish to have a larger return, but usually the goal is to lower the RCS.

In general it is desired to have the RCS lowered at all aspects and at all frequencies. (One might say that in many cases invisibility is the ultimate goal.) From a practical standpoint, however, the success of an RCS-control program is limited by constraints imposed by other disciplines involved in achieving the desired performance of the vehicle in question. That is, any suggested change in design aimed at lowering the RCS must be compatible with the performance criteria for the missile or aircraft.

Fortunately, RCS-reduction priorities can usually be established that make an RCS-reduction problem both feasible and practical. That is, some aspects usually are more important and some frequencies more crucial than others. For such situations one can often obtain a decided improvement in RCS characteristics via an RCS-control program.

Two basic methods are available for RCS reduction: (1) alteration of scattering configuration so as to redirect reflected energy from an aspect region of concern into other aspect regions of little or no concern; and (2) installation of radar-absorbing material (RAM) on a crucial scatterer. Before moving to a discussion of RCS reduction for some specific scatterers, it is worthwhile to note some general observations that can be made about the use of items 1 and 2.

1. When reduction is the RCS modification desired, it is obtainable only over a limited aspect region and then only at a cost of increased RCS in some other aspect region. Moreover, if success is achieved by gaining considerable reduction of RCS for a given component scatterer, then care must be exercised to check whether other component scatterers, previously considered to be insignificant contributors, are now producing a significant RCS. As the frequency becomes low enough that body length or wingspan is of the order of the wavelength, configuration changes become impossible. Thus there is a lower frequency limitation on this technique, imposed by the body size; but the more the frequency increases, the more useful the method becomes.

2. For absorber use in reducing RCS, there are also aspect and frequency limitations. First, the off-normal absorptivity performance of a RAM does

not usually match its normal incidence performance, falling off in direct proportion to an increase in angle of incidence off normal. Second, it is sometimes the bistatic performance of an absorber that should be of optimum quality, as, for example, is desirable in an inlet duct. Here again, it is generally the rule that as the size of the bistatic angle increases from 0° (normal incidence), the absorptivity performance of the RAM worsens. Here, as in item 1 above, difficulties attendant on RAM use increase as radar wavelength increases. It is generally true for all RAM that the lower the frequency for which the RAM is designed, the thicker, and usually heavier, the RAM must be. An exception to the increase in weight with an increase in thickness for RAM can be seen in the "Salisbury screen" absorber ([19] for a definition and description), which in fact can be made of material no heavier than a sheet of blotting paper located an appropriate distance from a metal surface; obviously, however, such RAM could have no applicability for any body that must move in the atmosphere. Frequency bandwidth over which a material can perform efficiently as an absorber is generally rather narrow if its thickness and weight are to be kept within reasonable bounds. Bandwidth may, however, be increased at an appropriate cost, such as the layering of materials designed to perform in different frequency bands, or the use of ceramic materials of increased complexity (the incorporation of magnetic properties), which is usually accompanied by increased weight and thickness. A detailed discussion of absorber characteristics and behavior has been given in [19].

Because of the weight penalty and frequency selectivity of RAM, the geometry-modification technique should be considered first, paticularly if there is some aspect priority involved in the RCS-control problem.

Each RCS-reduction problem is unique, but there are some general criteria that can be stated for component scatterers. However, as mentioned earlier, little can be done at HF frequencies. The RCS characteristics at such frequencies are relatively insensitive to changes in geometry, and a RAM designed for HF would in all likelihood be too thick and heavy for practical use. At higher frequencies (for moderately large bodies), however, practical RCS reductions can be obtained. To exemplify the problems that may be encountered and the sorts of solutions that may be proposed in seeking to reduce the RCS of an elongated, winged, and finned body propelled by engines requiring inlet and exhaust cavities, we shall discuss some RCS-reduction ideas for the most troublesome scattering components for such a vehicle: cavities (the next section); surface-wave supporters (traveling waves, the following section); nose and broadside scatterers; and wings and fins. We assume that interest extends over a wide frequency range but centers in a restricted aspect region about the nose-on direction.

RCS Reduction for a Duct

As was pointed out in equations 1.1 and 1.2 of chapter 1, when a cavity aperture area is in the vicinity of two square wavelengths (that is, $A/\lambda^2 \approx 2$; or $kd \approx 10$ if the aperture is circular of diameter d), the cavity on-axis RCS changes its behavior from flat-plate-like scattering given by $4\pi(A/\lambda^2)^2\lambda^2$ (or $0.05(kd)^3\lambda^2$) when $A/\lambda^2 < 2$ to a less sharply decreasing form determined empirically to be $17.8(A/\lambda^2)^{3/2}\lambda^2$ (or $0.05(kd)^3\lambda^2$ when $A/\lambda^2 > 2$. Thus, for wavelengths such that $A < 2\lambda^2$ there is no treatment available for the cavity interior that will affect the RCS because the energy does not enter the cavity to any extent for these area-to-square-wavelength ratios. Exterior treatments at the cavity mouth can, however, be profitably used for any value of the area-to-square-wavelength ratio. Therefore, these will be taken up first to illustrate the item 1 configuration-modification technique for RCS reduction.

Reflecting screens may be installed at a cavity mouth at an angle appropriate to reflect impinging energy to aspect regions of little or no concern, at the same time preventing the cavity from scattering energy into an apsect region of concern. The highest frequency of interest will govern the features of the screen that make it a reflector of electromagnetic energy (for all lower frequencies also) but allow it to pass air into the cavity. The simplest approach is to make the screen a flat-plate reflector, since then the direction of the largest reflected (and quite narrow) beam can be controlled precisely. But it is possible to give the screen a curved face, thereby lowering, but spreading out, the general level of its RCS return. As to the screen design, there are trade-offs available among reflectivity, wire radii, and mesh size. These trade-offs are shown in figure 1A–1 as a nomogram. Any of the parameters—wire size, mesh size, and energy transmitted—may be fixed and a range of available values determined from the remaining two ordinates. Or two values may be fixed and the third determined uniquely on the remaining ordinate. For example, if the frequency of interest is taken to be 17.5 GHz ($\lambda = 1.7$ cm) and if the radii of screen wires are taken to be approximately 0.5 mm spaced on a grid at intervals of 5 mm, then the power transmitted by the screen (one-way) is reduced by at least 10 dB, or 90 percent of the energy is reflected into some preplanned, unimportant direction.[1]

In lieu of the screen addition, there are other cavity modifications that can effectively reduce RCS at frequencies for which $A > 2\lambda^2$. In general, if the inlet throat itself can be bent or elongated in such a way as to prevent incident energy from striking directly on the compressor face, then the RCS will be appreciably lower than it would be if the inlet provided an uninterrupted conduit for incident energy to strike the compressor face. The amount of

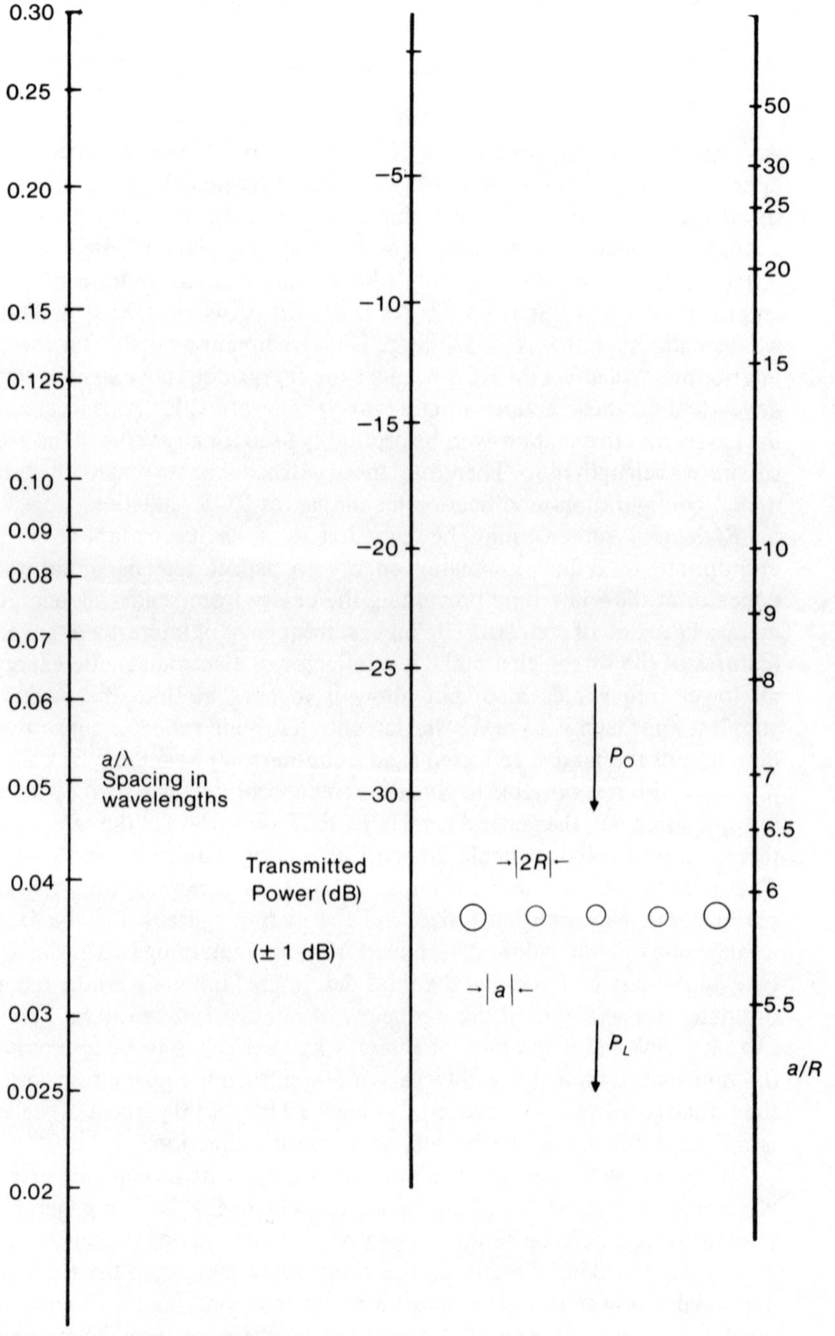

Source: Adapted from C.G. Bachman "A Study of Nanosecond Pulse Techniques in Radar Transmission," RADC-TR-61-247 (ASTIA 26880), 27 February 1961.

Figure 1A–1. Transmission through a Grid of Wires: A Nomograph

RCS reduction will depend on the actual shape of the duct itself; for the case of a rectangular throat in particular, an impinging plane wave will maintain an essentially plane front as it travels in the duct and may produce unwanted peak reflections at other off-axis angles.

If geometric modifications are not admissible, then RAM must be used to reduce RCS. Obviously, there are many levels of amounts of RAM that may be placed in a duct, and often a properly located partial coating will perform as effectively as a complete coating. The locations producing the greatest effectiveness are usually at any bend or throat constriction in the duct where the most bounces of energy may be intercepted.

The best possible scheme for reducing the RCS of an inlet duct is one that combines geometric modification with RAM application. Usually the geometric design can then be arranged in such a way that a complete RAM coating of the duct interior is unnecessary. However, as mentioned earlier, there are serious difficulties if the frequency range is wide, extending from frequency cutoff ($\lambda/2 < d < 3\lambda/2$) upward by as much as a factor of twenty.

To give a more concrete discussion of the foregoing principles, we have constructed in figure 1A–2 a generic duct model. it is assumed that the inlet orifice is flush with the underside of the vehicle. Assuming the aspect region of interest to be a conical one of 20° half-angle about the forward on-axis direction of the missile, we have in figure 1A–2(a) shown the location of a screen that would reflect energy downward at an angle of 25°. The screen size is sufficient that its main beam would be well outside the 20° cone of aspect interest for a frequency in the Ku band.

Figures 1A–2(b) and (c) show before an after concepts of the inlet duct. Inlet version (c) should have an on-axis (to the duct throat) RCS almost 10 dB lower than inlet version (b) at a Ku-band frequency. Figure 1A–3 shows before (b) and after (c) average RCS curves, as well as the peak that would be obtained from the screen. The desirable features to be stressed in version (c) are, first, that the inlet duct is formed so that almost all energy must bounce on the duct walls many times, both going in and coming out, and, second, that almost no energy can be directly incident on the compressor face. Even the energy that is so incident must undergo many bounces to get out of the duct. These aims are accomplished essentially by the elongation of the duct throat and by slanting and restricting somewhat the entryway into the inlet throat. Combinations of screen and redesign approaches incorporating these criteria are, of course, admissible.

RAM could be used with either version (b) or version (c), but with the latter it yields its greatest effects in combination with the configuration change. If RCS reductions are desired over a wide frequency range, say L through Ku bands, then very heavy requirements will be imposed on RAM choice, undoubtedly necessitating a layering of absorbers designed to

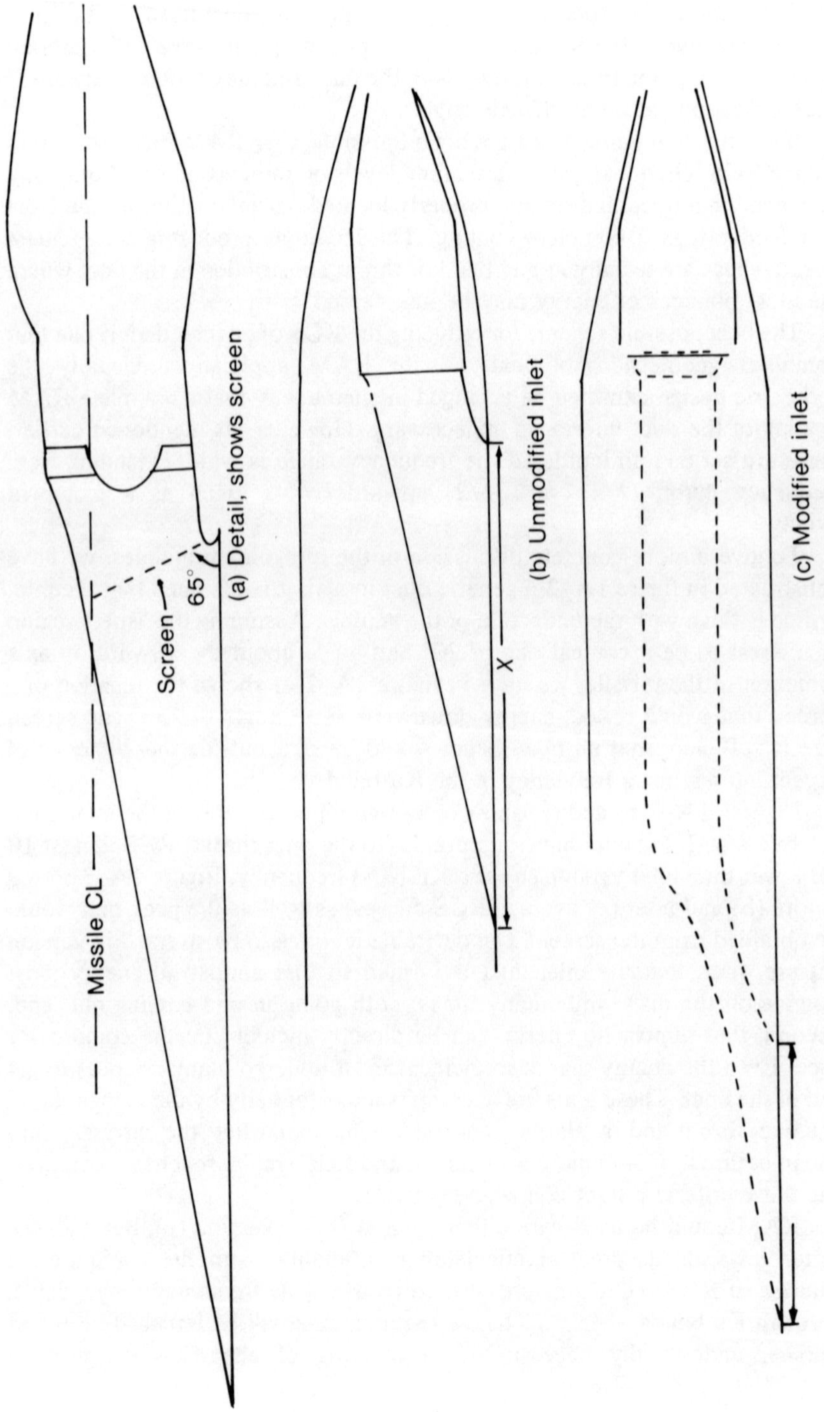

Figure 1A–2. Generic Inlet Model

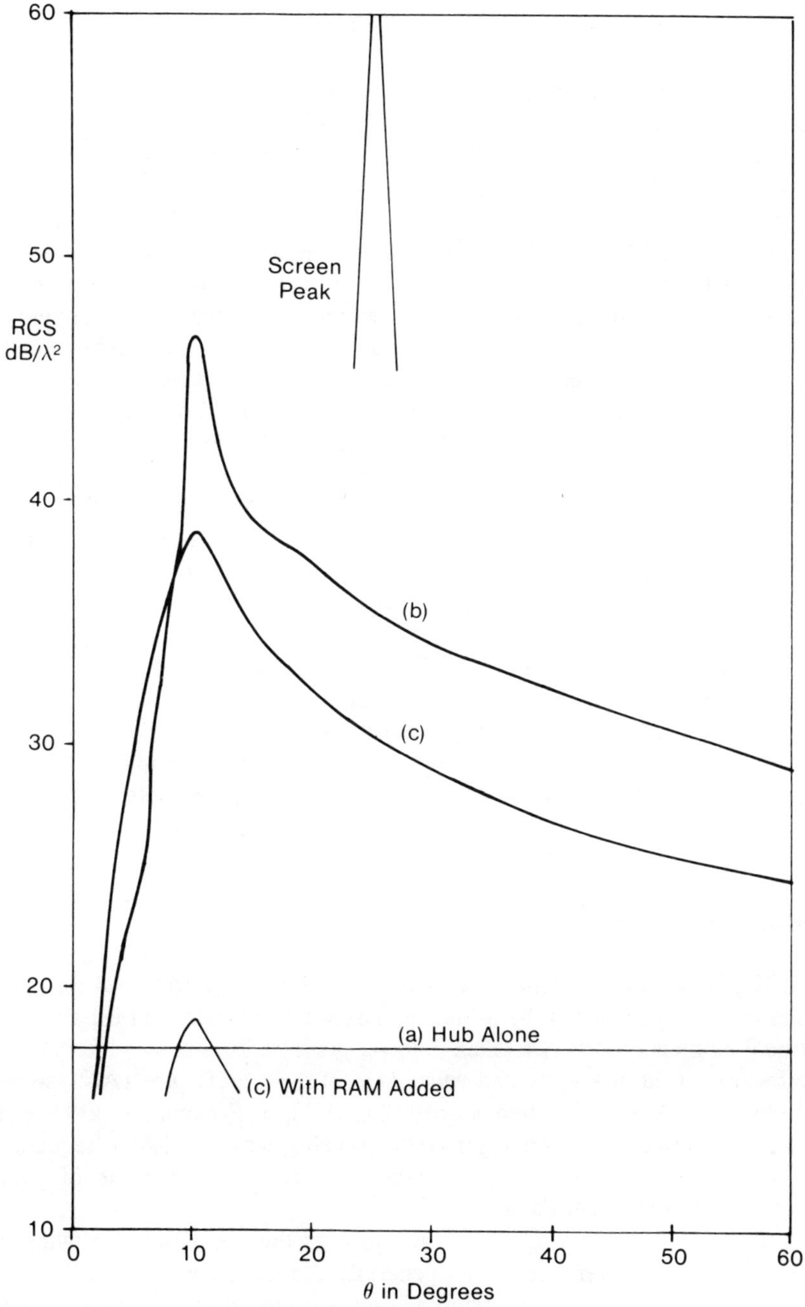

Figure 1A–3. RCS for Generic Inlet Drawings of Figure 1A-2 at a Ku Band Frequency

perform in separate frequency bands, or else a rather thick layer of ceramic absorber. For example, it is quite possible to design a ceramic (ferrite) absorber to perform well at S- and X-band frequencies; it will then, in general, *not* perform well at L- and C-band frequencies.

We will give two possible applications of RAM to the geometric variations of the generic model of figure 1A–2. First, consider (a): if a 10-dB RAM were applied to the hub and the duct throat forward of the compressor face, then the RCS would be reduced over the entire aspect region of interest by at least 10 dB.[2] This statement is true for the on-axis direction even though energy can strike directly on the compressor face; it is assumed that the compressor face configuration is such as to reflect all impinging energy toward the RAM-coated inlet wall. Thus all energy incident in the aspect region of interest will be reflected at least once from the RAM.

In (c), RAM could be placed on the horizontal portion of the duct wall (that is, from the compressor face forward to the bend) and on the hub (not shown). For such a situation, all energy entering the inlet would bounce at least twice on a RAM-coated surace yielding at least a 20-dB reduction of RCS, producing the lower of the curves shown in figure 1A–3—that is, a 20-dB reduction over and above the 10 dB gained by the geometric modification.

To sum up, then, we see that RCS reductions of 10–20 or even 30 dB are achievable for cavity scattering components, but at certain costs, such as covering the inlet throat with a wire mesh, elongating the inlet duct, or adding extra weight (and most likely other complications) in the form of RAM, all of which no doubt seriously affect other operating characteristics of the missile.

Reduction of Traveling-Wave RCS

RCS generated by traveling waves is essentially inconsequential for missile shapes at frequencies in the S-band and above. This statement is particularly true if no measures can be taken to lower the inlet RCS at these frequencies. However, if the RAM-treated version (c) of the inlet (figure 1A–3) were to be used, we would find that its resulting RCS, at S-band, would be of the same order as the RCS arising from the traveling wave. At lower frequencies, say, L-band and UHF, the traveling-wave RCS can become of greater concern than the inlet RCS.

If an absorber can be placed at and near the discontinuities where the traveling wave is reflected, then generally the traveling-wave RCS can be lowered by an amount approximately equal to the normal-incidence power absorptivity of the absorber. In the installation of such an absorber, care must be taken not to create a discontinuity (or impedance mismatch) as great as the one the absorber is supposed to eliminate. To avoid such an impedance mismatch, it is usually necessary to achieve a careful tapering of the absorber onto the metal surface.

If the traveling wave were of such significance as to require a treatment as drastic as coating the entire vehicle with RAM, then considerable RCS reduction can be gained, in fact much more than would be indicated by the normal-incidence power absorptivity of the absorber.

Two experimental examples will serve to illustrate this type of performance. A 10:1 prolate spheroid, 1 ft long, was completely coated with an absorber whose normal-incidence power absorptivity at 1.3 GHz was only 2 dB and at 3 GHz was just 12 dB. When the spheroid was measured at these frequencies, its broadside RCS was reduced by 2 and 12 dB at 1.3 and 3 GHz, respectively, as would be expected. But the traveling-wave RCS of the spheroid was reduced by 11 and 22 dB at these two respective frequencies. This "remarkable" performance of the absorber in the reduction of the traveling-wave RCS is due to the fact that the absorber had the opportunity to attenuate the amplitudes of the surface currents as they traveled over the length of the spheroid.

Without the RAM treatment, the relative importance of the traveling peak in the RCS pattern of the prolate spheroid at the two frequencies was quite different. At the higher frequency, where the spheroid-length-to-wavelength ratio was 3.08, the traveling-wave peak was 11 dB below the broadside peak. At the lower frequency however, where the spheroid-length-to-wavelength ratio was only 1.35, the traveling-wave peak was only 2 dB below the broadside peak.

It does not follow from the preceding two paragraphs that significant RCS reductions can be obtained for the traveling-wave contribution of a missile at HF frequencies. We emphasize, therefore, that the experimental results referred to were obtained at relatively short wavelengths, even though the spheroid-length-to-wavelength ratios were small, since the experiments were carried out on models. At HF frequencies the absorber required for a missile would be prohibitively thick (and no doubt prohibitively heavy as well).

At S-band and higher frequencies the traveling wave becomes among the least significant of the component contributors to the RCS of a missile. In particular, if $kL > 75$, then the envelope of σ/λ^2 remains relatively insensitive to changes in other parameters governing traveling-wave RCS, providing the relative phase velocity is held at 0.98 (within the envelope the peak locations will vary with λ and L, L being the missile length), and takes a maximum value of 3 at approximately 11° off-nose. Thus, for example, a 200″ missile at 3 GHz ($kL \cong 125$) could produce a maximum traveling-wave RCS no greater than -15 dB/m^2.

RCS Reduction for the Body

In most cases the body (or fuselage) of the target is troublesome (from the RCS-control point of view) as a source of a traveling wave or because of the broadside flash. On occasion the nose configuration may also yield a return that is too large for small θ. The problem presented by traveling waves was discussed in the preceding section; here we shall attempt to consider general techniques for the other aspects of the body problem. Of course, one can retain the geometry (which yields the too-large return) and use RAM, but this usually involves a large weight penalty. Therefore, let us concentrate on geometry changes.

If near-nose-on aspects are of prime concern, then improvement in RCS performance can often be obtained by pointing the nose; that is, an ogival front section is preferable to a spheroidal one. A comparison of RCS characteristics is given in figure 1A–4, where for illustrative purposes we have assumed that the body diameter is 21 inches and that the spheroidal nose is a 3.5:1 prolate spheroid. From the figure we see that for this case and for $0° \leq \theta \leq 30°$ one could obtain an RCS reduction of 20 dB at all frequencies above 3.0 GHz. At lower frequencies the RCS reduction would be less, since the average RCS for the ogive varies like λ^2, whereas that of the spheroid varies like λ^0.

It should be emphasized that the foregoing assumes the nose to be metallic. If there is a radome in that location, one is presented with an entirely different problem, one for which band-pass radome materials would probably be required.

The fuselage invariably presents its largest return at or near broadside. Then $\theta = 90°$, the fuselage presents a return similar to that of a cylinder; and for small λ we have the optics return, kaL^2. It may be that for the frequencies of concern this large peak lobe in the pattern will be sufficiently narrow (as measured in θ) so as to be of *relatively* little concern. In that case one may decide to tolerate the large RCS at $\theta = 90°$.

If something must be done about the large broadside response, then the next step is to determine whether some large responses at $\theta = 90°$ can be tolerated if they occur only at isolated roll angles. If that is the case, then a charge in the fuselage from a circular cylinder to a rectangular cylinder may be of benefit. This geometry change is illustrated in figure 1A–5. For the special case of $x = 3.00''$, $y = 1.50''$ $(= 2a)$ and $\lambda = 0.39''$ (with $L = 6.75''$ and $r/\lambda = 0.2$), we have RCS comparison displayed in figure 1A–6.

From a theoretical analysis the RCS of the circular cylinder, σ_1, and the RCS of a square cylinder, σ_2, should have the ratio $\sigma_2/\sigma_1 = 8a/\lambda$ at $\phi = 90°$ (that is, 12 dB compared with the 8–10 dB observed in the figure). At $\phi = 135°$, theory predicts a ratio of $\sigma_2/\sigma_1 \approx 0.57(2a/\lambda)^{-1}$, assuming a "sharp" edge (−8 dB compared with the observed −11 dB in the figure).

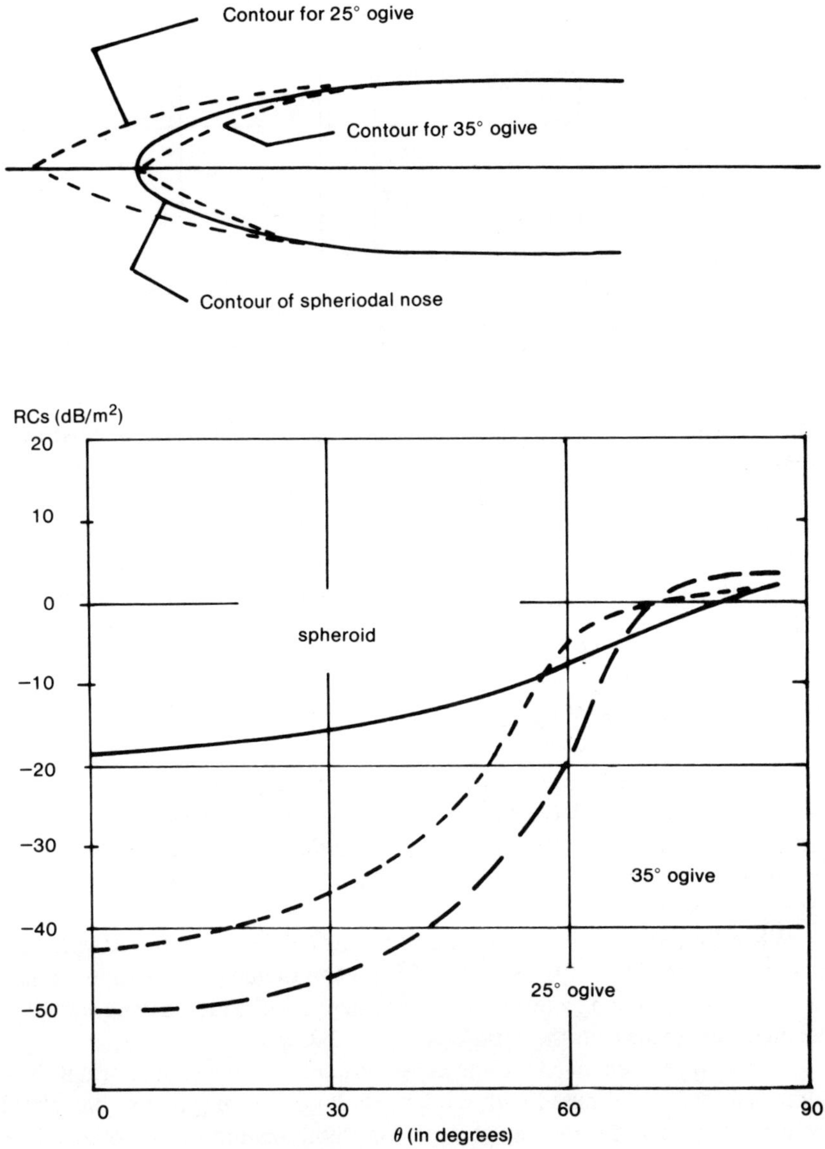

Figure 1A–4. Possible Modification in the Design of the Nose Section of the Body (with RCS Values at 3.0 GHz)

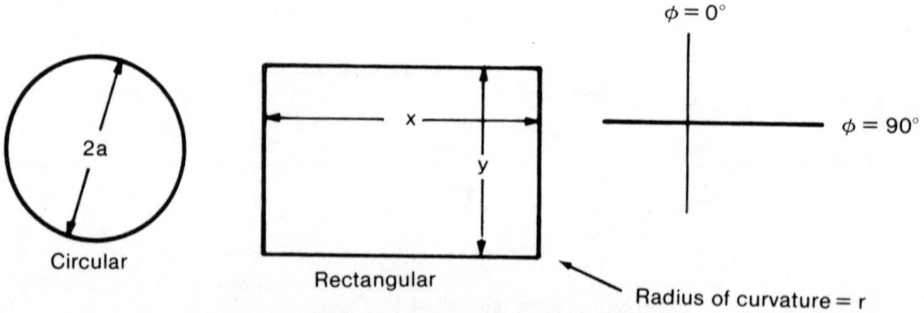

Figure 1A–5. Fuselage Possibilities

Thus, to within about 3 dB, theory and experiment agree for this illustrative case.

It thus follows that for high frequencies (where $2a/\lambda$ would be large) squaring off the fuselage would result in an RCS "improvement" if there were four aspects (say $\phi = 0°, \pm 90°$, and $180°$) at which one could tolerate an *increase* in RCS in order to obtain RCS reductions for all other roll angles.

RCS Reduction for Wings and Fins

The wing and fin surfaces present an RCS problem, particularly at the higher frequencies, only at isolated aspects. One must have an aspect that is normal to the leading edge or to the trailing edge in order to obtain an RCS value of significance (or be normal to the flat surface itself). Here we assume these edges to be straight.

When the aspect is in the plane of the wing (or fin), then the RCS (at the higher frequencies) is of the order of L^2/π when the edge is sharp, or it may be like a cylinder giving a return of $k\rho L^2$ (where ρ is the radius of curvature of the edge) depending on the ratio ρ/λ.

A special problem occurs if the aspect is confined to a plane normal to an edge; then one is presented with a relatively large return (greater than L^2/π) for almost all aspects in that plane at one polarization or the other. This occurs, for example, with regard to a trailing edge (in particular), which is normal to the body axis, and in the yz plane when θ is measured in the $\phi = 0°$ plane.

We can illustrate this by considering a square flat plate of edge length $2a$ placed in the xy plane and considering its RCS as a function of θ and ϕ via physical optics. Assume that $2a/\lambda = 5$; then the RCS as a function of θ and ϕ will be as shown in figure 1A–7. It is clear from the figure that if there is

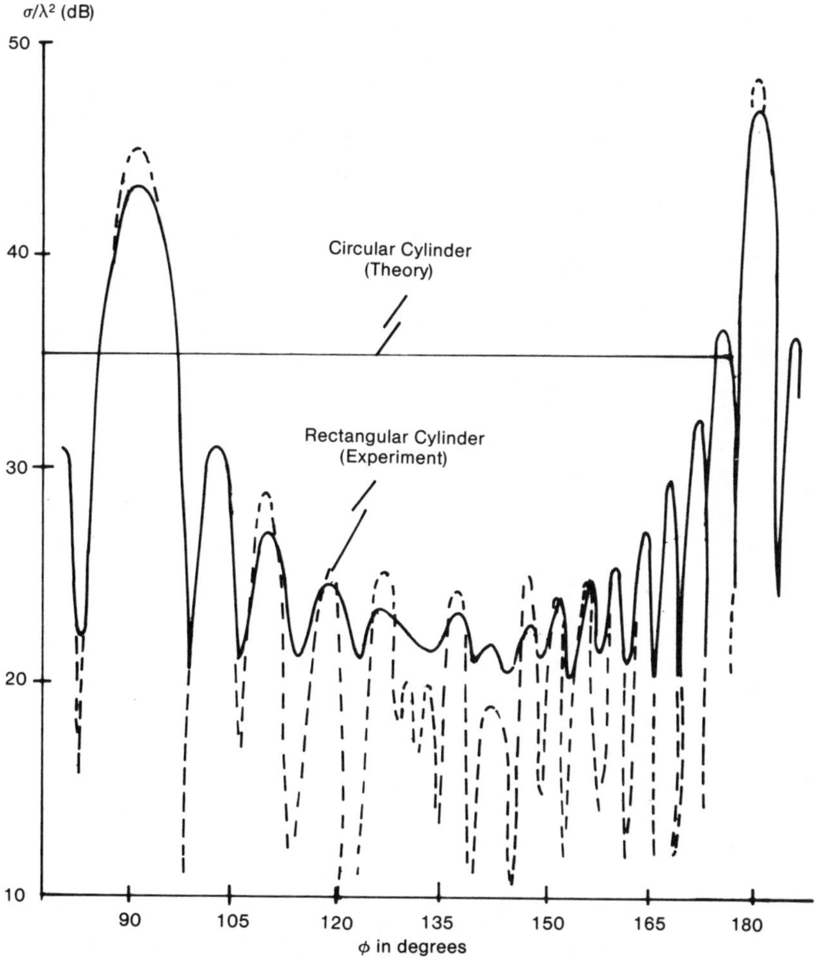

Note: From figure 1A–5: $x = 3.0''$, $y = 1.5''$, $\lambda = 0.39''$.

Figure 1A–6. Comparison of Fuselage-Type RCS Values

a plane of aspects of special concern (such as those passing through the axis of the missile or aircraft fuselage), then one should avoid having the fin or wing edges normal to that axis.

In the foregoing we have assumed these edges to be straight. Some RCS reduction can be gained by curving the edges. Figure 1A–8 illustrates the concept for a wing. This modification would reduce the amplitude of the peak, although it would spread out the reduced peak over a wider θ range.

A second approach (possibly applicable to the fins, but not to the wings)

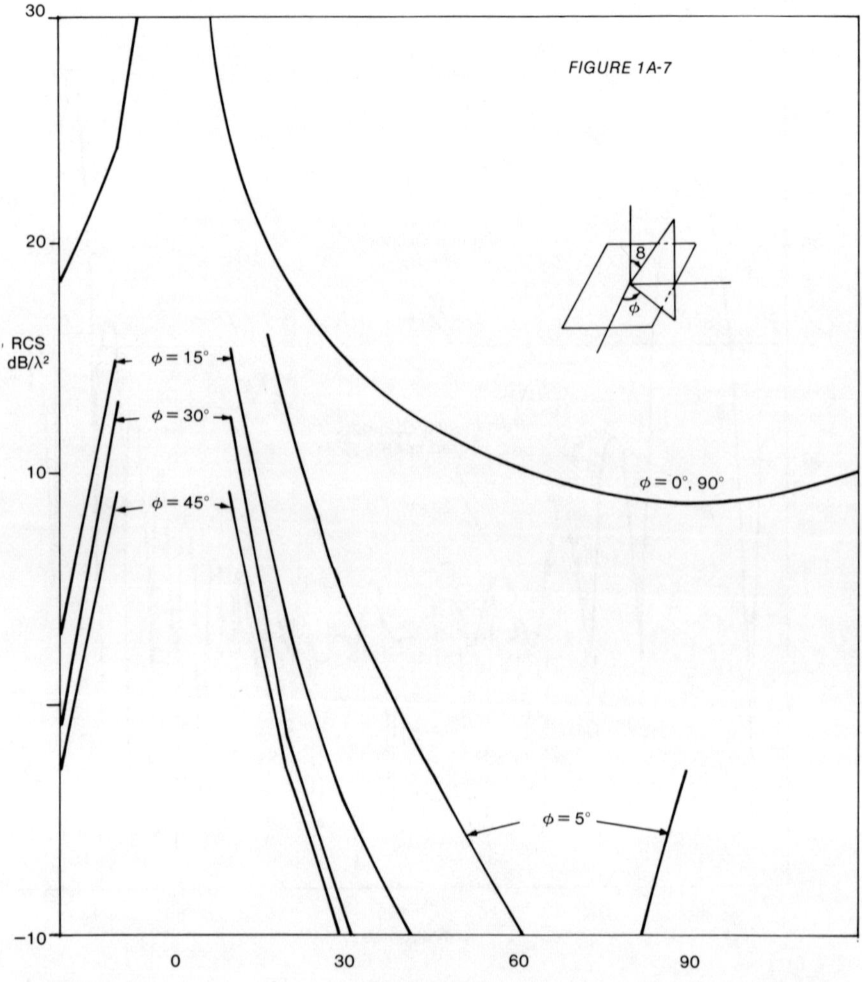

Figure 1A–7. RCS of a Square Flat Plate 5λ on a Side (Physical Optics)

would be to curve the plates constituting the fins. Figure 1A–9 illustrates this geometric concept for the nose-on view. In general it is desirable to avoid having any of the fin edges normal to the axis of the missile. Figure 1A–10 gives a generic illustration of what we consider bad and good fin-edge arrangements.

It has been assumed in the foregoing discussion of the wings and fins that their structure is largely, or altogether, of metal (see figure 1A–8). If it is possible to form these members entirely from fiberglass, then it is possible in

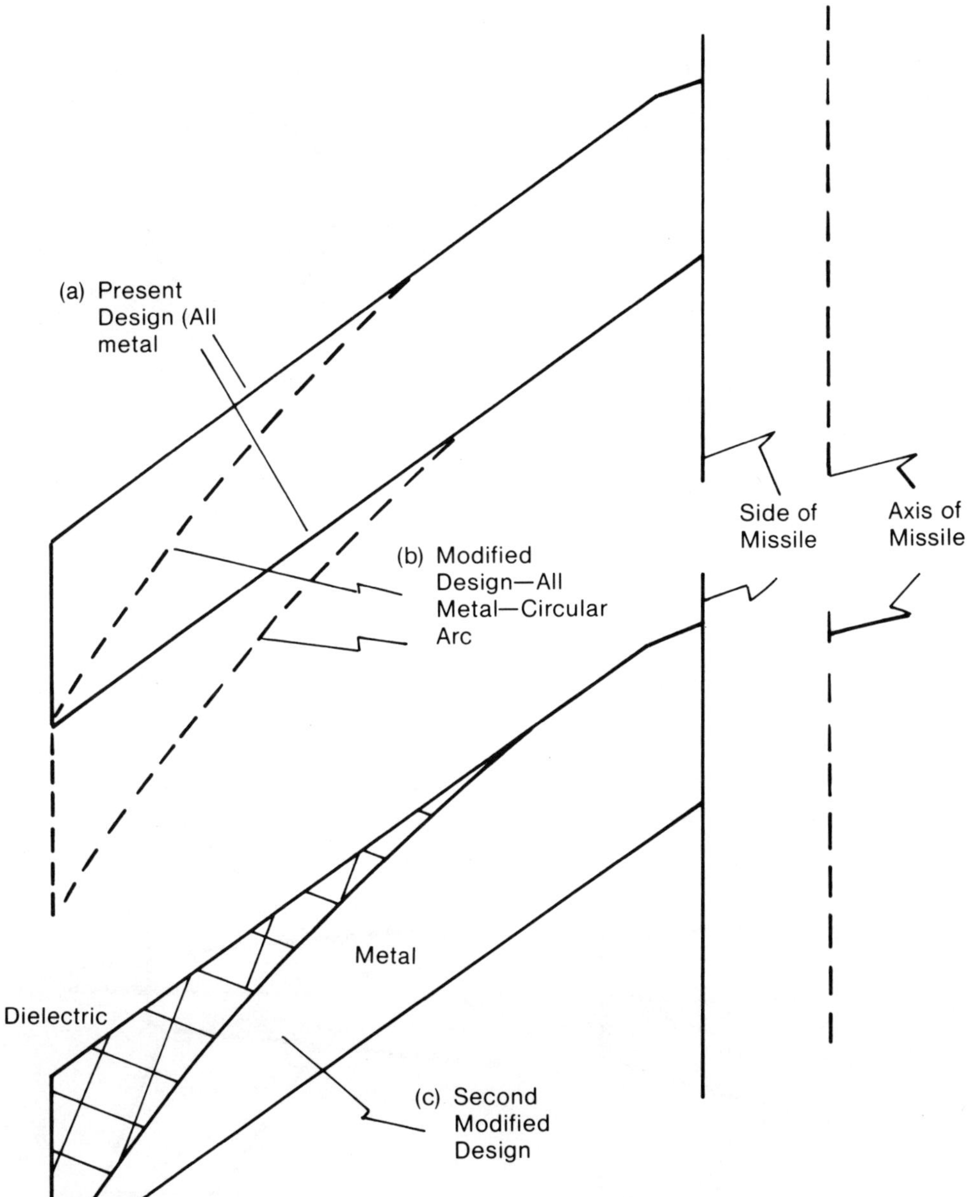

Figure 1A–8. Possible Modifications in the Wing Design

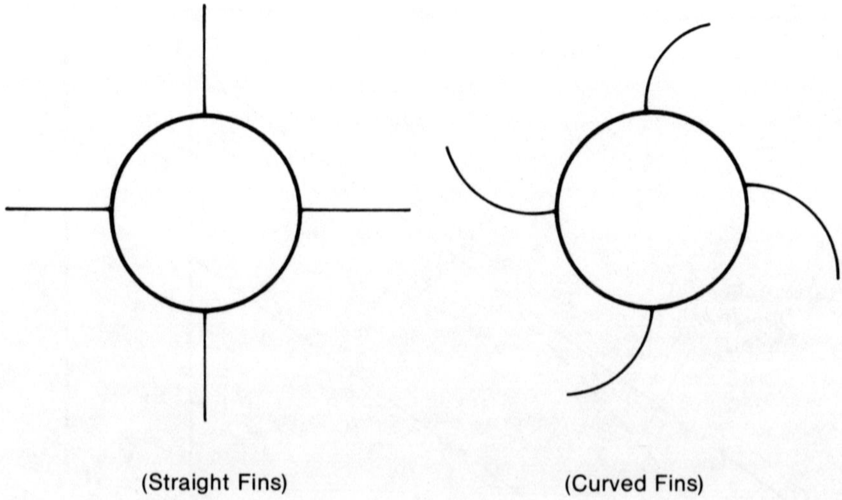

Figure 1A–9. Possible Fin Configuration

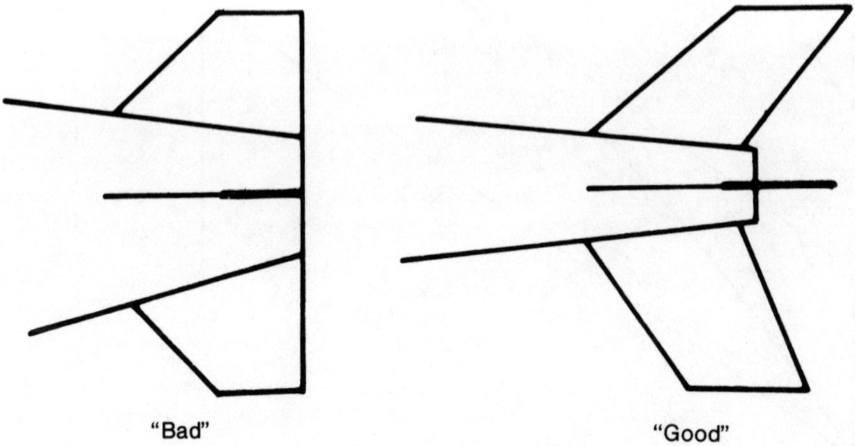

Figure 1A–10. Fin Designs

general to realize a reduction in RCS. The size of the reduction can be estimated by applying the flat-face (power) reflection coefficient, where the radius of curvature is much larger than the wavelength, to the RCS determined for the metal scatterer. This reflection coefficient is

$$|R|^2 = \left|\frac{1-n}{1+n}\right|^2,$$

n being the index of refraction of the fiberglass. If we assume that the material has no magnetic properties so that $\mu_r = 1$, then

$$n = \sqrt{\varepsilon_r}$$

(n is actually $= \sqrt{\varepsilon_r \mu_r}$, where μ_r is the relative permeability. Taking an index of refraction for fiberglass as low as 2, we find that $|R|^2 = 1/9$, so that the RCS for the fiberglass member would be approximately 9.5 dB lower than the RCS for the corresponding metal member.

Radar-Absorbent Materials

Materials designed to absorb incident electromagnetic energy are called *radar-absorbent materials* or RAM. Common commercial uses have been in measurement facilities for obtaining antenna patterns and radar-target reflectivities to reduce, or essentially eliminate, extraneous reflections that could cause errors. Military applications have been to reduce radar return of aerospace or seaborne vehicles and ships, making radar detection and tracking for weapons guidance difficult.

RAM shapes have ranged from combinations of sharp pyramids and flat surfaces to absorbing particles embedded in dielectric materials. Ambient operating environments, weights, volumes, costs, and effects on sensors operating at other wavelengths are all considered in particular designs. Compromises between operation over a range of frequencies for varying incident angles and reduction magnitude of radar return are often necessary. in some applications, bistatic reflectivity characteristics are important.

The operation of RAM applied over a conducting surface is shown in figure 1A–11. When the incident wave meets the conducting surface, it is reflected. If an absorber is placed over the conducting surface, partial reflection takes place. Ideally, the surface impedance of the absorber over all incident angles would be identical to free space, and no reflection would occur. Physical properties of known and available materials preclude this possibility, and design compromise is necessary. To function properly, the

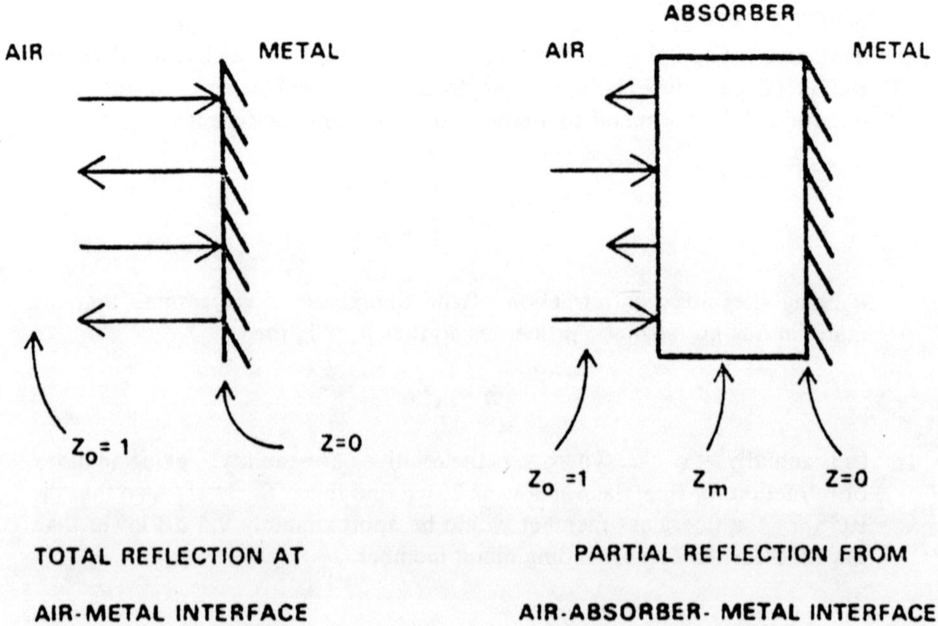

Figure 1A–11. The General RAM Problem

RAM must match the impedance closely *and* absorb the incident energy. Deviations from these two conditions over the operating frequency range and the range of incident angles cause partial reflection.

RAM materials are sometimes evaluated by plotting energy reduction in dB in the ordinate of a Cartesian graph for two orthogonal abscissas, each pointing in opposite directions—one over the operating frequency range, the other over the range of incident angles.

In general, conformal panels of RAM placed on the exterior of a radar target have a thickness related to the lowest incident frequency needed to be effective. This thickness is approximately one-tenth to one-quarter of the wavelength corresponding to the lowest frequency. The target surface area, the thickness required, and the radar-absorbing material's density then determine the required weight penalty.

To illustrate the effects of frequency on reflectivity from materials of different thicknesses. Figure 1A–12 illustrates data from several fiberglass laminate sheets with a relative dielectric constant of approximately 4, as compared with a metal plate. Three thickness values are shown. Over much of the 1–18-GHz range, the fiberglass supposed to be transparent to microwaves is only 4.5 dB less than reflectance from a metal plate. Deep

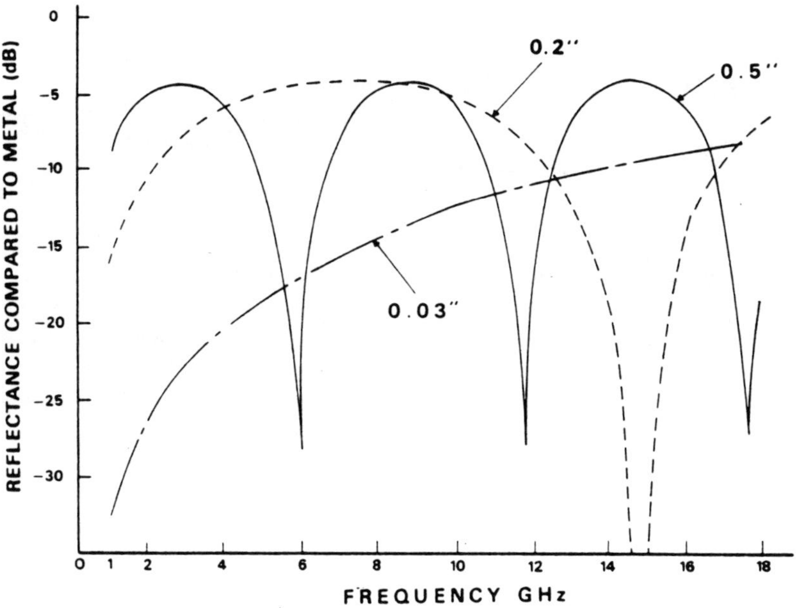

Figure 1A–12. Fiberglass Laminate Reflection

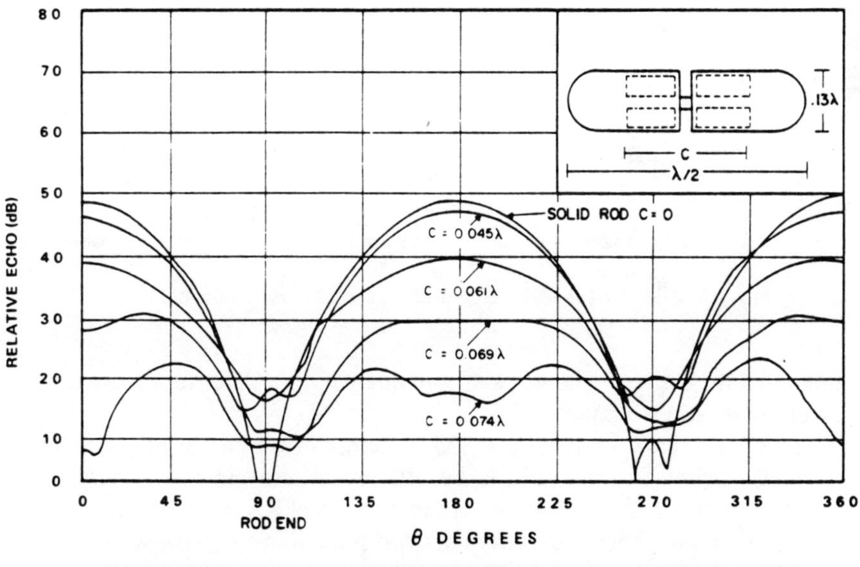

BACKSCATTERING CROSS SECTION VS TARGET ANGLE FOR VARIOUS LOADINGS OF A SINGLE LOADED ROD. POLARIZATION, HORIZONTAL. FREQUENCY = 3100.8 MHz

Figure 1A–13. Impedance-Loading Techniques

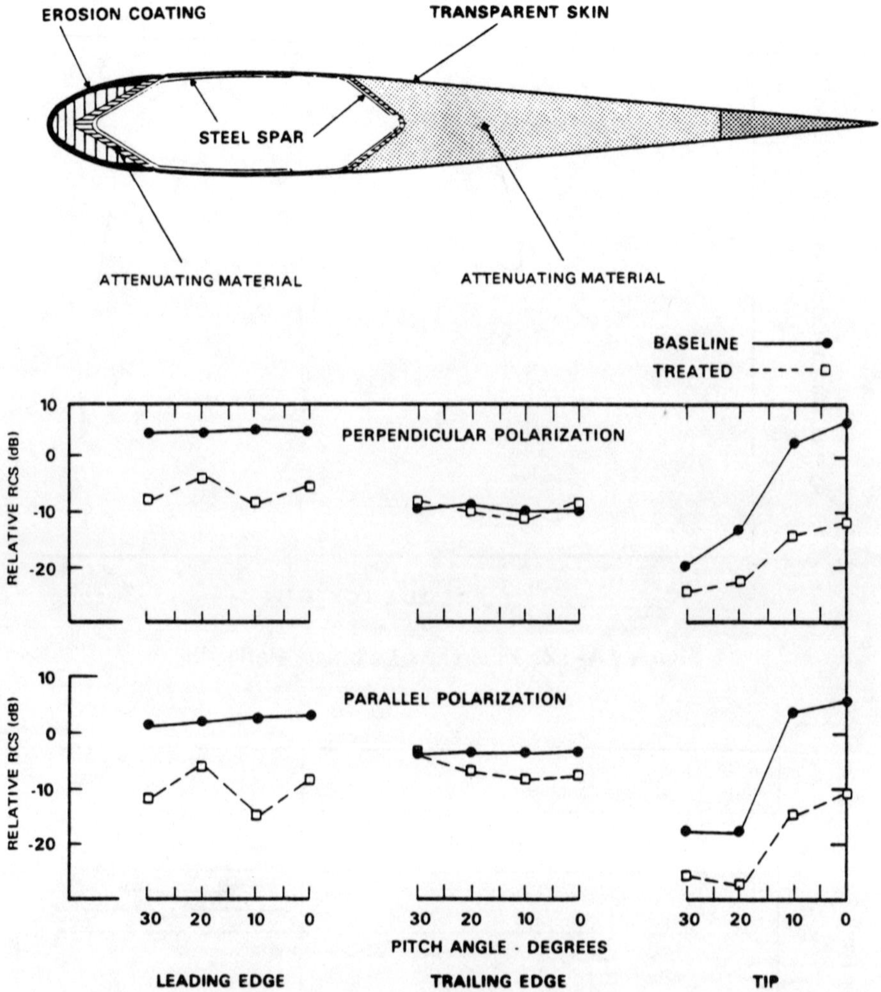

Figure 1A–14. Rotor-Blade Echo Reduction

nulls in the curves shown can provide design data for radomes, or electromagnetic windows.

It is customary to specify RAM in terms of its bandwidth, maximum coverage angle of normal incidence, dB reduction in incident energy, and weight of unit per area. When required thickness of RAM become prohibitive for applications, other radar-cross-section reduction methods are available. One such method is to provide impedance loading for a target class of rotational symmetry, as shown in figure 1A–13. A reentrant cavity with dimensions related to incident wavelength is placed at the center of the cylindrical target to absorb incident broadside of energy. Since resonant effects are exploited in the design, it is narrow band and extremely frequency sensitive.

Typical results of radar-cross-section reduction by RAM application are shown in figure 1A–14. Both baseline and treated reflectivity data are illustrated to show effects of orthogonal polarizations over a small range of incident angles from various aspects. Between one and two orders of magnitude reflectivity reduction were obtained by RAM treatment of this configuration.

Radar-cross-section reduction techniques using RAM often combine shaping or alteration of the vehicle configuration to provide an inherently low radar cross shape, and the use of dielectric lenses to achieve an overall composite design meeting desired specification goals. Often aerodynamic, thermal, structural-cost, and materials-availability constraints limit design options. Observables resulting from these designs to other sensors such as visual, infrared imagers, or laser radars can also have an impact on design choices.

Notes

1. In lieu of a screen it is also possible to install an "egg crate" in the duct throat. The end-on view of the egg crate would have dimensions and reflectivity properties like those of the screen. It may, however, offer advantageous structural or air-ducting features superior to those offered by a plane screen.

2. An absorber with 25-dB absorptivity performance at normal incidence may yield only 18 dB at 30° incidence and 5 dB at 70° incidence; it may perform even less effectively for some bistatic situations. Thus, to account for all incidence situations, we assume an *effective* absorptivity level of 10 dB even though the RAM may be rated at 25 dB.

2 Survey of Passive Enhancement Techniques

Introduction

It is often desirable to simulate the radar response of a large, complex target by the use of a small, inexpensive substitute. Applications that immediately come to mind are those involving drone or towed targets for testing complete systems. Although the final test of any such system is how well it detects a real target, it is obvious that we must rely on substitutes of one sort or another.

Fortunately, we don not need to rely entirely on guesswork to predict the radar response of a target. We can construct a model, using available data, and measure the response of this scaled-down model at an appropriate scaled-up radar frequency. This type of simulation has been used to predict detection ranges and other gross characteristics of the radar response of operational targets. The term *gross characteristics of the radar response* is used not because the model technique is incapable of yielding the fine structure of the radar response of a target (its variation with frequency, polarization, and target aspect), but because the practical application of these data requires a detailed knowledge of how the radar "uses" the echo signal to track the target, how fast the target closes, the dynamic response of the antenna servo system, and a host of other factors peculiar to each radar system.

A more practical use of the model data is to design a substitue target that can be flown or towed, is small and inexpensive, and has the same radar response as the original aircraft. Using this target, we can come very close to an operational test of any complete defense system. To translate the measured radar response of a target into the design of a much smaller substitute, one must devise enhancement techniques, which are essentially techniques for making small things look big to a radar system. Fortunately, radar systems are easily fooled because they measure size in a very special way.

The designer of a substitute target employs as elements of that target certain basic reflectors that have large radar cross-section (or apparent size). This chapter will discuss some of the most common reflector elements, their limitations, and how several have been modified to meet new enhancement problems. We shall also discuss possible applications of enhancement techniques outside the field of simulation.

Basic Enhancement Devices

There are several requirements for a good passive enhancement device. First, it must be efficiently excited by the incident radar pulse. Second, it must concentrate the reradiated energy in the direction of the receiver. Since most radar systems locate source and receiver together, we are primarily interested in backscattering, or the so-called *monostatic* case. Third, it should operate in this manner over a considerable range of attitudes to the radar, which we will call the *coverage region*. Last, it should have as small a volume as possible consistent with the first three objectives.

Two types of enhancement device can be distinguished. One may be called *optical*, since its operation can best be explained by the methods of optics; the other category contains those devices whose mode of operation is more nearly that of an *antenna*. Optical methods are especially useful at high frequencies, where optical elements such as lenses and mirrors can be employed to focus and direct reflected energy efficiently. At intermediate and lower frequencies, we may employ antennas, which are essentially devices for efficiently intercepting or radiating waves along preferred directions.

Optical Devices

Characteristics of typical optical enhancement devices are given in figure 2–1. This list is by no means complete but illustrates those devices with broad coverage. It is apparent that the spherical Luneberg lens excels in this respect. Common features of these reflectors are as follows:

1. The elative echo pattern is broad and independent of frequency, provided the characteristic dimension of the reflector is greater than 3 or 4 wavelengths.
2. The peak echo area occurs when the reflector is broadside, presenting the greatest silhouette area. It varies as the square of the frequency and as the fourth power of the characteristic dimension.
3. If radar is linearly polarized, the reflected wave has the same direction of polarization as the incident wave. If the radar is circularly polarized, the reflected wave has the opposite sense of rotation to the incident wave.
4. Required reflector volume increases as the 3/4 power of the average echo area, for a given frequency and reflector shape.

The chief disadvantage of the optical reflector is the fixed volume requirement and the broadside characteristic of its response. At frequencies below 3,000 MHz it is difficult to find room for optical reflectors with a high ratio of surface area to volume. Wings, for example, cannot efficiently be

Type		Peak Echo Area	Principal Coverage Region	Average Echo Area (all aspects)
Square Corner Reflector		$\frac{12 \pi l^4}{\lambda^2}$	Cone of 23° total included angle about symmetry axis (to 3 db down points)	$\frac{0.7 l^4}{\lambda^2}$
Circular Corner Reflector		$\frac{15.6 l^4}{\lambda^2}$	Cone of 32° total included angle about symmetry axis (to 3 db down points)	$\frac{0.47 l^4}{\lambda^2}$
Triangular Corner Reflector		$\frac{4 \pi l^4}{3 \lambda^2}$	Cone of 40° total included angle about symmetry axis (to 3 db down points)	$\frac{0.17 l^4}{\lambda^2}$
Spherical Luneberg Lens Reflector	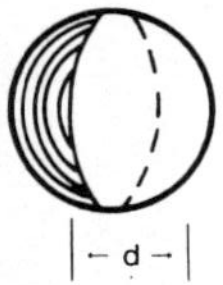 Conductor	$\sim \frac{2 \pi d^4}{\lambda^2}$	Cone of approximately 150° total included angle about symmetry axis (to 3 db down points) appreciable return over 180° cone	$\sim \frac{2 d^4}{\lambda^2}$

Figure 2–1. Optical Reflectors

used to house optical reflectors; and the echo area required at the nose aspect of the target may be higher than a broadside reflector with the same silhouette area will yield. There are ways of trading bandwidth for volume and transverse dimensions for longitudinal extend (end-fire operation), providing we use antenna concepts in designing a reflector.

Antennas as Reflectors

We can illustrate the behavior of antennas as reflectors by presenting radiation and measured reflection patterns for a polyrod antenna [1]. These are shown in figure 2–2. The echo area pattern is to a close approximation the square of the antenna power pattern, with the peak echo area given by

$$\sigma_{\text{peak}} = \frac{G^2}{4\pi}\lambda^2$$
$$\sigma_{\text{peak}} = 225\lambda^2, \tag{2.1}$$

where G is the power gain of the polyrod as a radiator. The same peak echo area could be obtained with a broadside reflector such as a flat plate with an area of $4.2\lambda^2$. Note that this would require 17 times as much transverse cross-sectional area as the polyrod. We have illustrated how to trade transverse cross-section for longitudinal extent.

Antennas that are characterized by a single radiating mode exhibit reflection patterns that are just the square of the antenna power pattern, and peak echo areas given by equation 2.1. More than a single radiating mode exists, each can contribute to the reflecting pattern. A good example of such an antenna is the sectoral horn, which has a reflection pattern quite unlike its radiating pattern, because of higher-order modes that are set up in the horn by the incident wave but are not excited in the antenna mode.

Another application of the antenna concept, suggested by Van Atta, uses coupled antennas to obtain the equivalent of an optical corner reflector without using as much volume. Figure 2–3 shows the principle for a linear array. Corresponding antennas on either side of the center of the array are connected through transmission lines of constant length. It is found that a wave incident on the array from an arbitrary direction gives rise to an array mode with a maximum in the direction of the source. We can reduce volume requirements in this manner, but only at the expense of bandwidth.

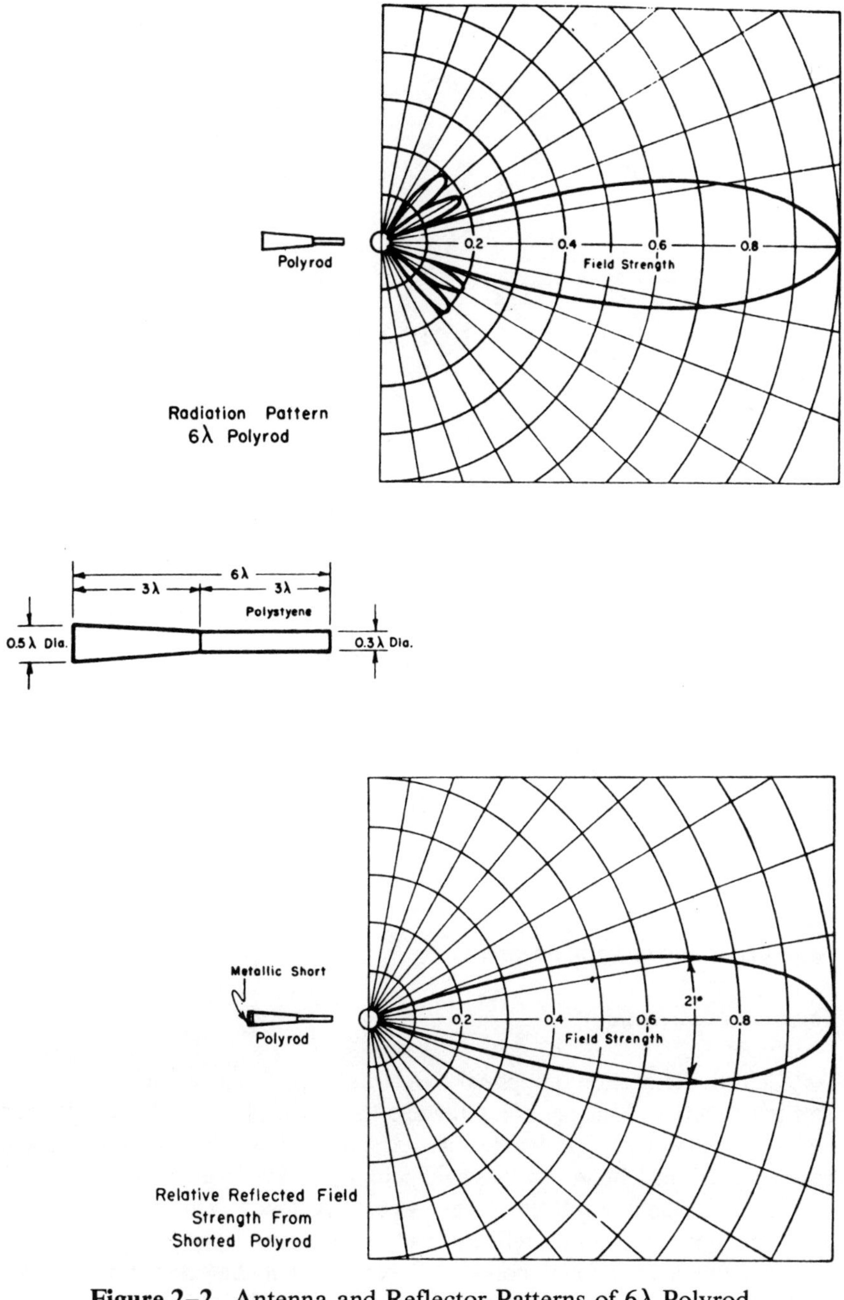

Figure 2–2. Antenna and Reflector Patterns of 6λ Polyrod

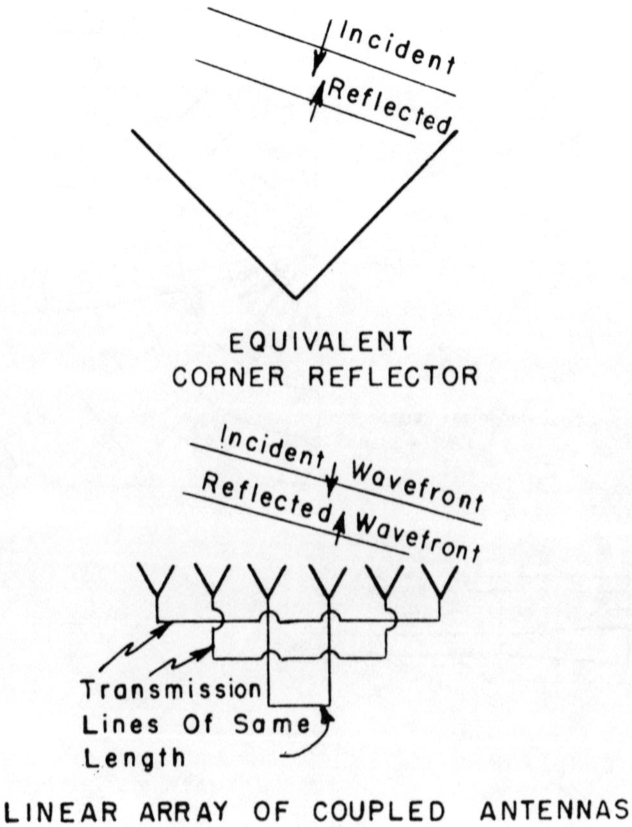

Figure 2–3. Van Atta Array

New Problems in Enhancement

Circular Polarization

As noted in the section on optical reflectors, triple-bounce corner reflectors and the spherical Luneberg lens reverse the sense of a circularly polarized wave. A radar using the same circular polarization for transmitting and receiving will be "blind" to these reflectors [2]. Two means of overcoming this defect in corner reflectors have been developed. The first involves a minor change in the corner reflector and is frequency dependent. The second replaces the conventional corner reflector with a dielectric filled corner having two conducting walls and one dielectric wall. This reflector works well over an extremely wide frequency band.

Dielectric Sheet Modification of Corner Reflector [1]. If we insert a plane slab of dielectric material parallel to one face of a corner reflector, as shown in figure 2–4, maintaining the slab thickness and air gap as shown, the circularly polarized response (to a radar using the same circular polarization for transmitting and receiving) will be within 2 dB of that given in figure 2–1. The bandwidth of the layer shown is approximately ±10 percent of the design frequency. The linearly polarized response of this corner will not be affected by the dielectric slab, except for a very narrow region of aspects.

Dielectric-Filled Reflector [1]. If we construct a reflector as shown in figure 2–5 of a low-loss dielectric material with $2.3 \leq \varepsilon_r \leq 4$, we will obtain circularly polarized response within 1.5 dB of that given in figure 2–1, over all frequencies for which the ordinary corner-reflector equation applies.

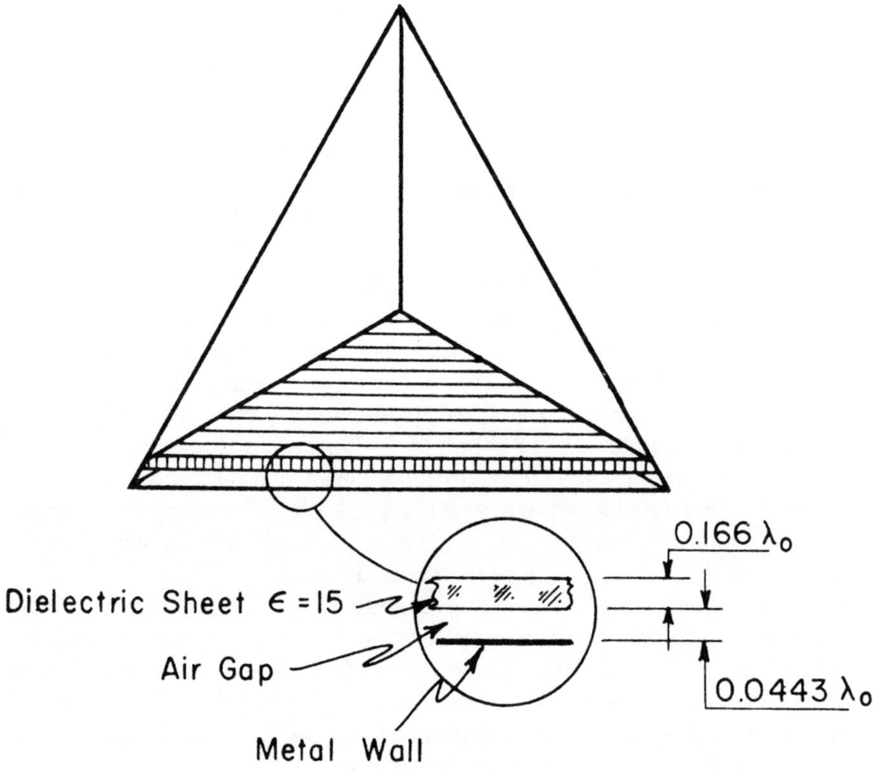

Figure 2–4. Dielectric Sheet Modification of Corner Reflector

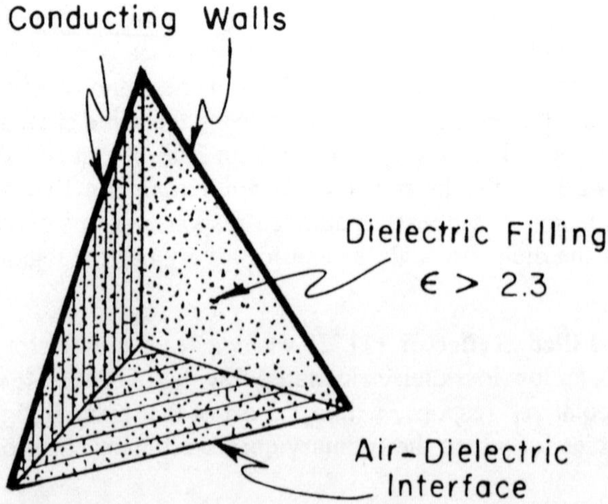

Figure 2–5. Dielectric-Filled Corner Reflector with Two Conducting Walls

Bistatic Enhancement

Another enhancement problem arises when source and receiver are not located together. This so-called bistatic configuration occurs in some missile systems, when the transmitter is located in the mother ship and the receiver is carried by the missile. After firing the missile, an angle of as much as 10–15° may accrue between transmitter and receiver lines of sight to the target. The conventional optical reflectors of figure 2–1 obtain large monostatic cross-sections at the expense of bistatic response, since the reflected energy is concentrated in a narrow beam directed back toward the source. Figure 2–6 illustrates a technique used to obtain limited bistatic response from a corner reflector. It is found that a single error in the wall angles of a corner reflector causes the reradiated beam to spread in a single plane, as shown in the figure. This spreading can be controlled by the error angle and is fairly constant over the principal coverage region of the corner reflector.

Figures 2–7 and 2–8 illustrate the effect of bistatic angle γ on the measured echo patterns of a normal corner reflector and on one with an error of 5° in one wall angle. Because the degradation of the reradiated pattern is confined to a single plane, the monostatic echo area is not greatly reduced by the error, and it is possible to achieve reasonable monostatic and bistatic cross-section with a cluster of these reflectors.

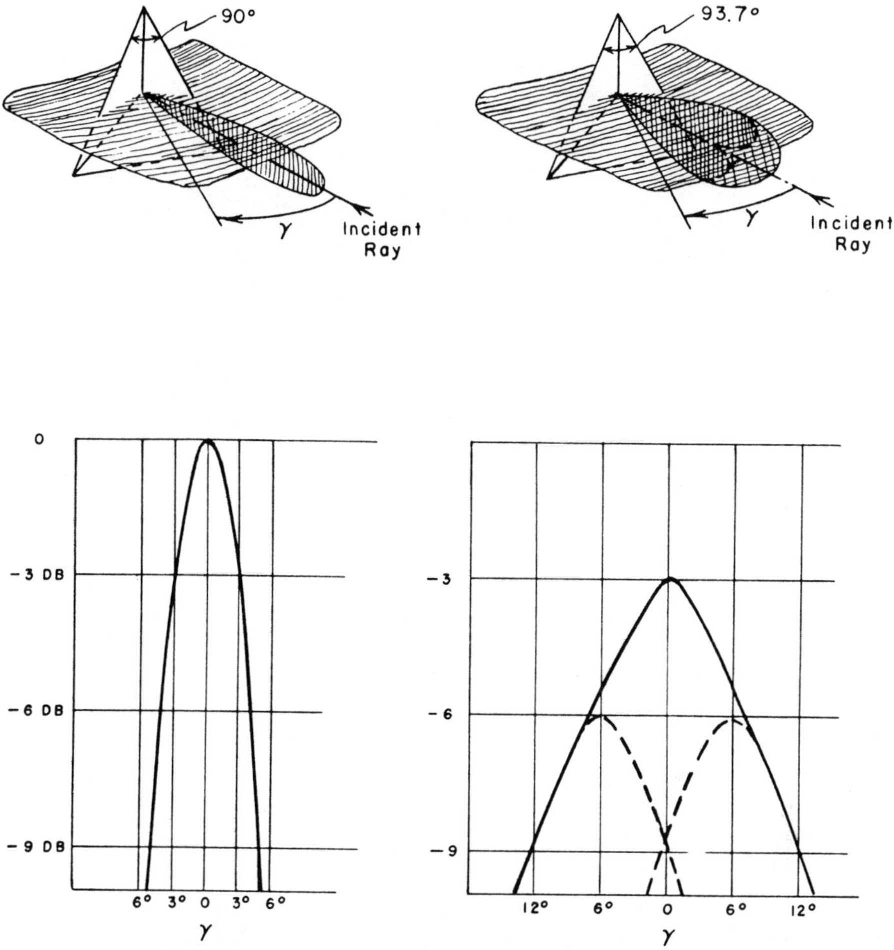

(a) Normal Corner (b) Coner with 3.7° Error

Figure 2–6. Effect of Error Angle on Reradiated-Beam Shape

New Applications for Enhancement Devices

Use of antennas as reflectors enables us to control enhancement, since the echo area depends on the terminating impedance. There are several methods for varying the terminating impedance electrically, that require very little power. By this means we should be able to impart a characteristic

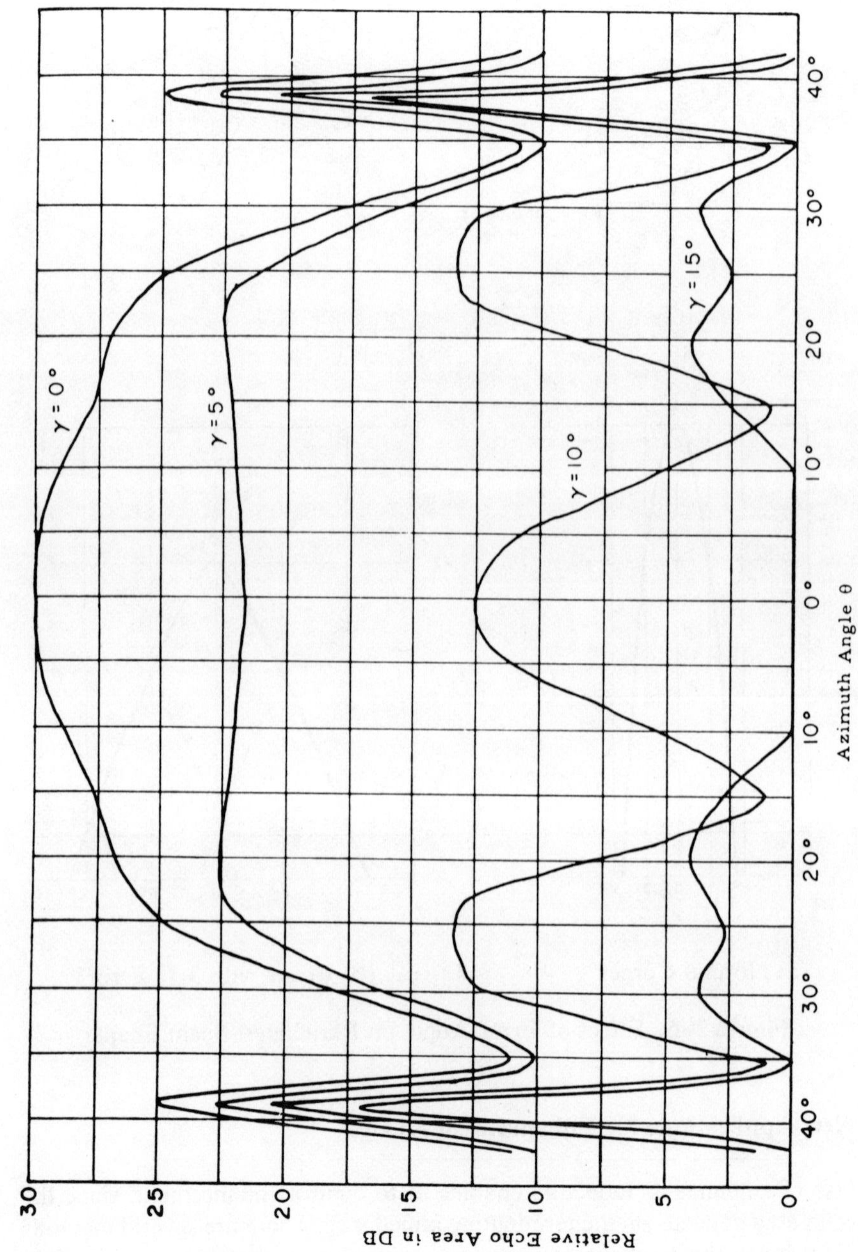

Figure 2–7. Effect of Bistatic Angle on Response of Normal Corner Reflector

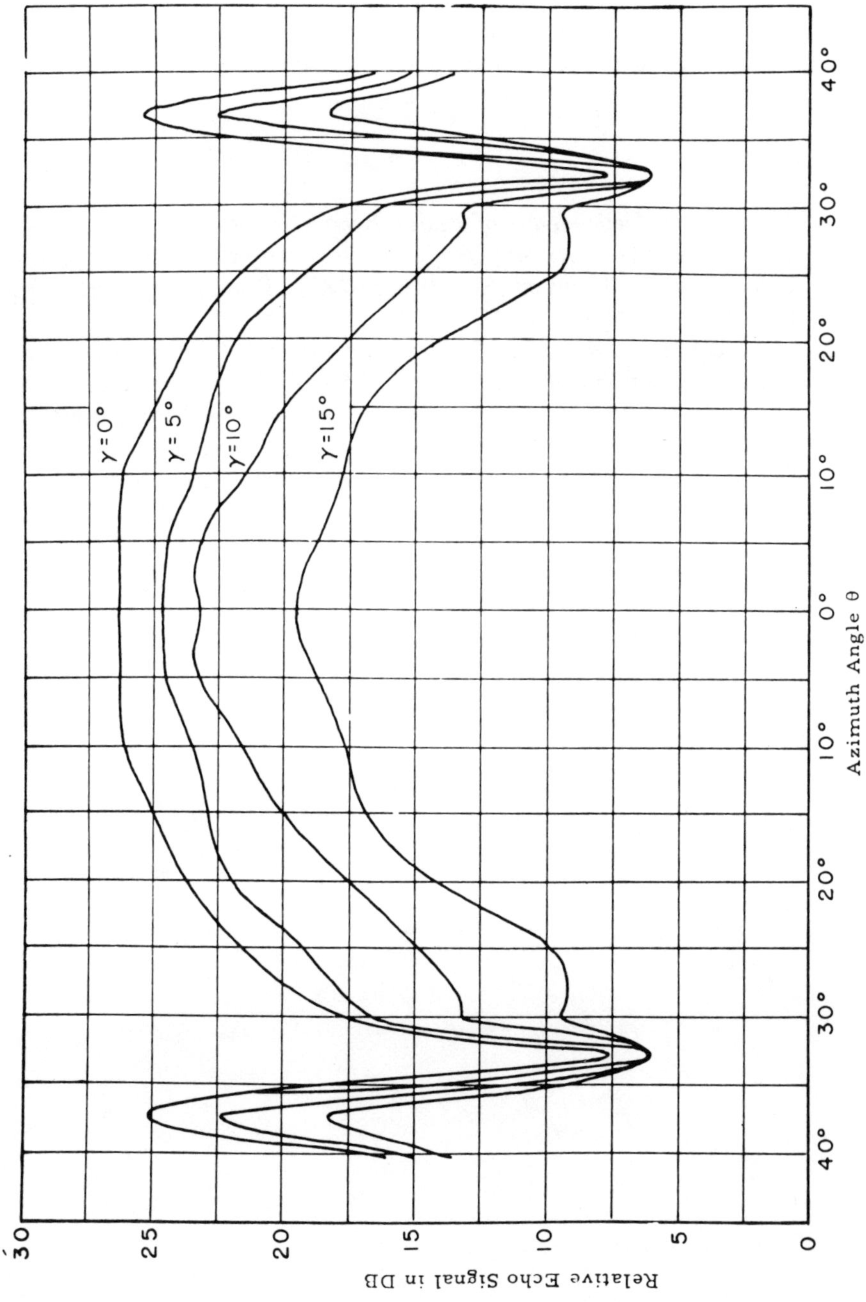

Figure 2–8. Effect of Bistatic Angle on Response of Corner Reflector with 5° Error

modulation to the echo signal from friendly aircraft, for identification purposes. In addition, amplitude and angle noise can be introduced into the echo signal, if desired, to produce large tracking errors.

References

1. *Polarization Dependence of Radar Echoes.* Report 612-4, 1 October 1955, Antenna Laboratory, Ohio State University Research Foundation. Prepared under Contract AF 30(635)-2811, Rome Air Development Center, Rome, N.Y.
2. Kennaugh, E.M. *A Corner Reflector for Use with Circularly Polarized Radars.* Report 601-8, 30 January 1956, Antenna Laboratory, Ohio State University Research Foundation. Prepared under Contract AF 33(616)-2546, Air Research and Development Command, Wright Air Development Center, Wright-Patterson Air Force Base, Ohio.

3 Spatial Coverage of Radar Reflectors

Summary

There is great difficulty in evaluating the spatial lobe structure of even simple radar reflectors by examination of standard backscatter patterns. When data are presented in the form of contour charts drawn on a suitable grid base, however, the three-dimensional coverage of different designs can be evaluated in a straightforward manner. The superiority of a circular corner reflector over square or triangular designs is demonstrated by plotting their backscattering areas in this manner. Comparison can also be made of altogether different types of reflectors, as illustrated by a jack, which provides very poor return relative to the three types of corners (table 3–1).

Introduction

Measurements of radar backscattering are generally presented as a set of patterns involving the two angles of the spherical coordinate system (figure 3–1) that may be erected about a model. In general, each pattern of the set is for a constant value of one of these angles, whereas the other angle varies over its complete range. Theoretical investigations often use this same pair of angles; but other parameters, such as direction cosines for corner reflectors, also provide simple functional expressions and are employed in graphical presentations despite their poor relation to physically significant quantities. Whatever the choice, it is difficult to visualize the complete three-dimensional lobe structure in space by examining the variation of backscattering in the conventional graphical presentations. It is possible, however,

Table 3–1
Performance of Reflectors

Type of Reflector	*Half-Power Angle*	*Probability that* $\sigma > 0.5\ \sigma_{max}$ (%)	*Probability that* $\sigma > 0.1\ \sigma_{max}$ (%)
Square corner	25°	7	37
Triangular corner	38°	23	58
Circular corner	46°	29	75
5λ jack	—	3	26

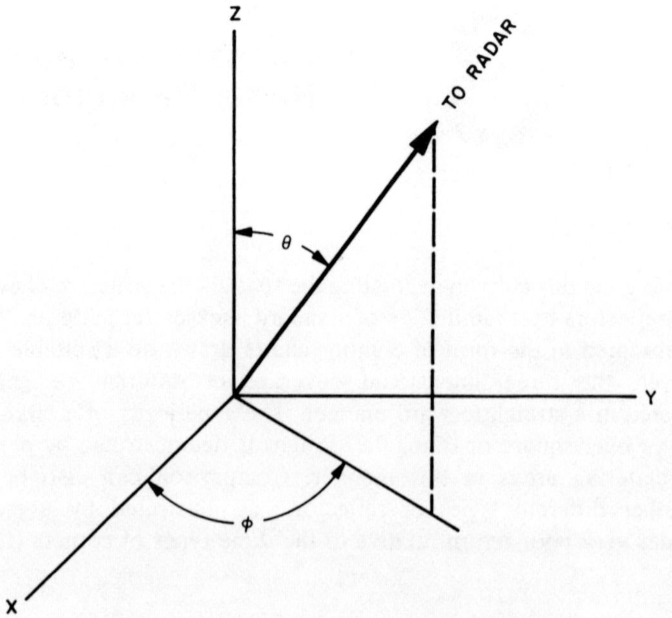

Figure 3–1. Spherical Coordinate System

to plot contours directly on a grid of spatially significant angles, as in the work of Kennaugh [1], who plotted the equivalent reflecting area of aircraft on the grid of the Lambert azimuthal map projection. This and other conventional map projections suitable for plotting radar-cross-section contours are described in this chapter.

Square-Corner Reflector

The square-corner reflector (figure 3–2) has been analyzed by Spencer [2], who expressed his theoretical analysis of the cross-section σ in terms of direction cosines l, m, and n, as follows:

$$\sigma = \frac{64\,\pi}{\lambda^2}\,\frac{l^2 m^2 b^4}{n^2} \qquad m \le \frac{n}{2},$$

$$\sigma = \frac{4\,\pi}{\lambda^2}\, l^2 \left(4 - \frac{n}{m}\right)^{k^2} b^4 \qquad m \ge \frac{n}{2},$$

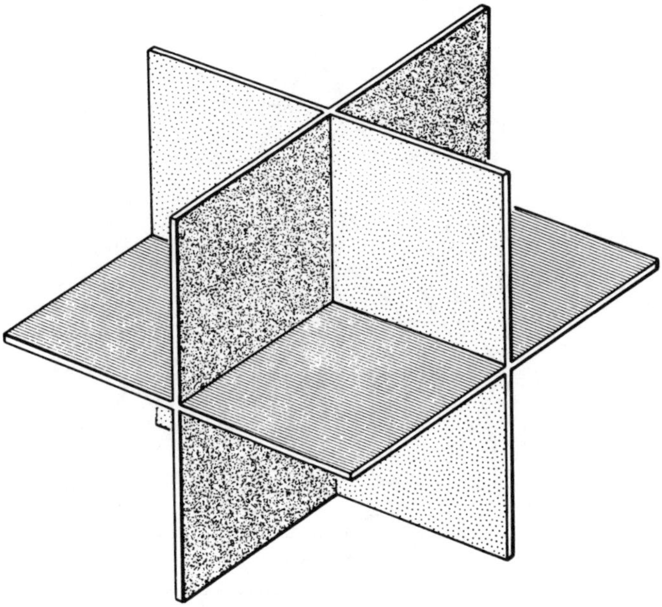

Figure 3–2. Cluster of Square-Corner Reflectors

where λ is the wavelength and b the length of one side of the square. The angular variation of return may be presented as contours of uniform level of radar cross-section on a direction cosine grid, as in figure 3–3, but the spatial variation of reflectivity is not readily intepreted in this coordinate system. The maximum backscattering is along the radial line that is equidistant from all three vertices, of the corner, as illustrated in figure 3–4. The value of the maximum is plotted in figure 3–5 for selected edge dimensions, b. The lines are dashed in the region in which the wavelength exceeds the edge dimension, since the derivation is based on physical optics and is therefore not applicable in this region.

With the value of the maximum return established, the next question is how rapidly the return falls off from its peak value. Before describing a contour chart, we orient the spherical coordinate system of figure 3–1 so that the coordinate axes are coincident with the edges of the corner, θ is the polar angle (or the colatitude) and ϕ the azimuthal angle (corresponding to the longitude of an earth coordinate system). It is appropriate to mention the latitude and longitude because the coordinate grids for the data presentation have been developed by cartographers for representation of the earth's surface.

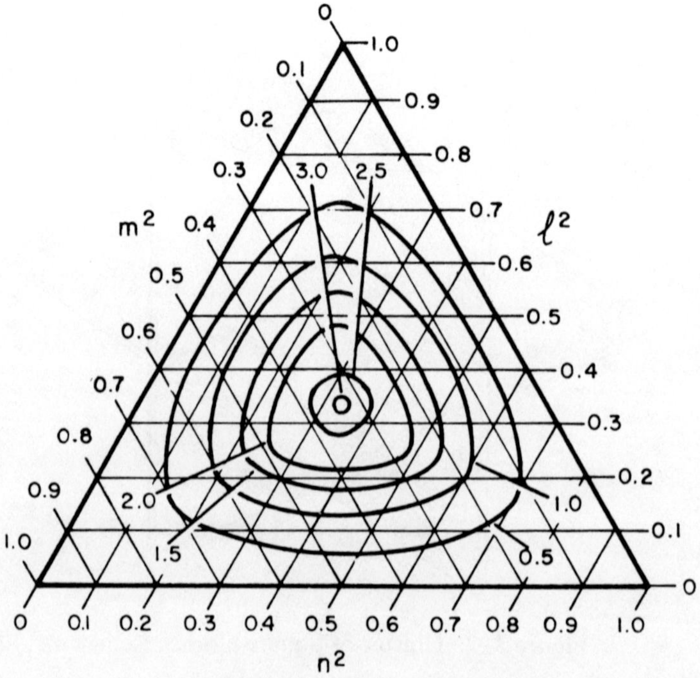

Source: K.M. Siegel, M.L. Barasch, J.W. Crispin, R.F. Goodrich, A.H. Halpin, A.L. Maffett, W.C. Orthwein, C.E. Schenstad, and C.J. titus, "Studies in Radar Cross-Sections XVIII—Airborne Passive Measures and Countermeasures," University of Michigan, January 1956, Appendix D, Figure C-5, p. 285.

Note: The contour values are expressed in units of $\sigma\lambda^2/4\pi b^4$.

Figure 3–3. Radar Cross-Section of Square-Corner Reflector

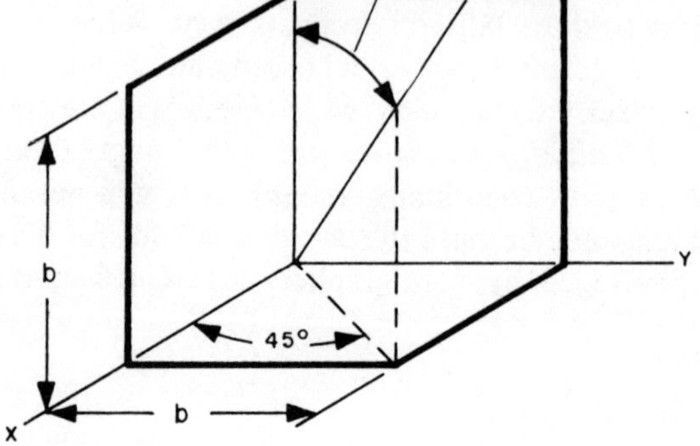

Figure 3–4. Direction of Maximum Cross-Section

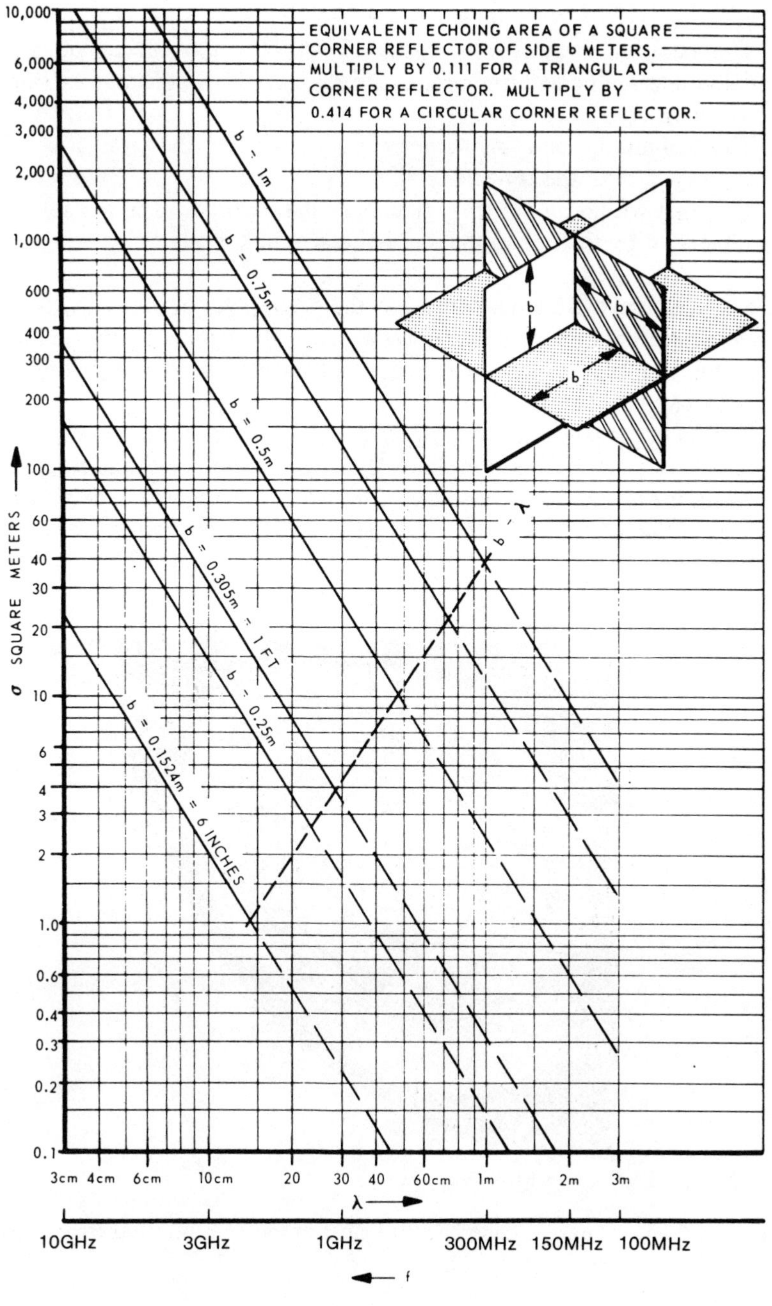

Figure 3–5. Maximum Cross-Section

The Lambert azimuthal projection furnishes a grid of θ and ϕ that has the following advantageous features:

1. The coordinate system is easily drawn, for the curves of constant colatitude are concentric circles and the curves of constant azimuth are radial straight lines.
2. The colatitude and azimuth of points within the grid are readily associated with the reflector.
3. The area within an arbitrary boundary is proportional to the solid angle subtended by the corresponding boundary in space.

As a consequence of the last property, the probability of observing any level of backscattering may be determined with a high degree of precision merely by running a planimeter over a contour, when the angular limits of a rendezvous problem are specified.

The theoretical equivalent reflecting area of a square corner is plotted on the Lambert azimuthal grid in figure 3–6, where the colatitude θ represents the angle measured from one edge of the corner (along the z axis), and the angle ϕ is the azimuthal angle measured in the perpendicular plane (the xy

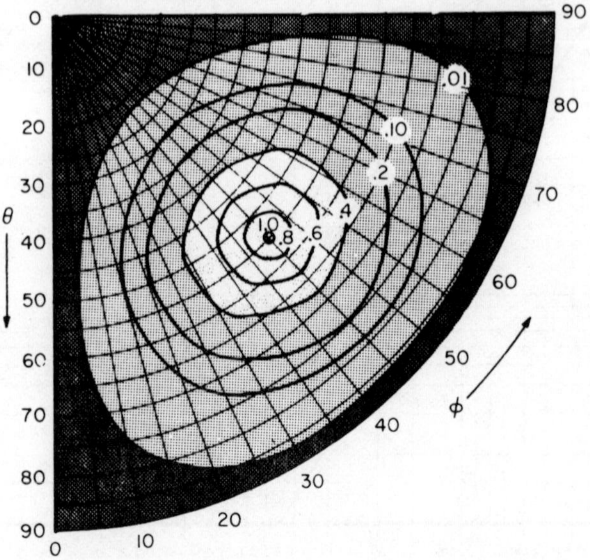

Figure 3–6. Square-Corner Return on Lambert Azimuthal Projection

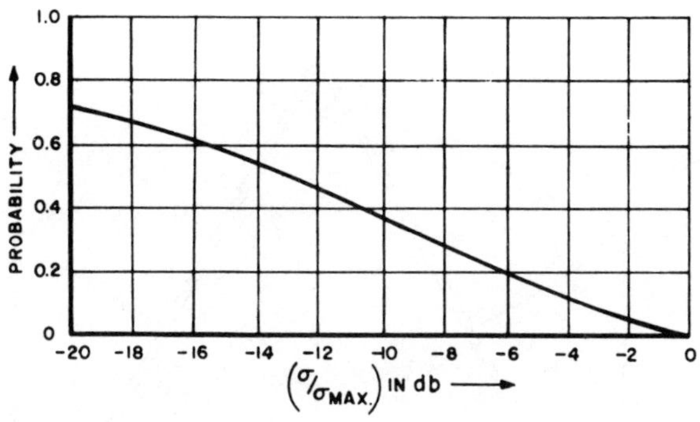

Figure 3–7. Cumulative Distribution for Square-Corner Return

plane). The contour plot shows very strong return centered about the peak, with quite a rapid falloff, and represents a rather highly directed reflectivity pattern. The reflecting area is within 50 percent of its peak value (analogous to the half-power beamwidth of an antenna) over an angle of 25°.

By measuring the area within each contour level, we obtain the probability of observing any level of return when the corner is viewed at a random angle, as long as the radar is within the octant where all corner surfaces are visible. The cumulative probability plot is given in figure 3–7, which shows that there is a 7-percent probability of being within 3 dB of the peak value and a 37 percent probability of being within 10 dB of the peak.

Triangular-Corner Reflector

Because of the narrow lobe of the square corner, Spencer [2] recommended use of the triangular corner (figure 3–8), whose broader coverage is plotted in figure 3–9. The reflecting area for this shape is within 50 percent of its peak value over an angle of 38°. For comparison with the theoretical analysis, we have plotted Robertson's measurements [4] (figure 10) on the Lambert azimuthal grid (figure 3–11). The dihedral returns, which show up as pronounced ears on the patterns of figure 3–10, do not extend over a great enough solid angle to be significant. Comparison of the theoretical results with measured values is readily accomplished by means of the cumulative probability distributions in figure 3–12. The two curves are quite close, with the measured values falling below the theoretical result. The perfect reflector, when viewed at random, furnishes return within 3 dB of the peak 23 percent of the time and is within 10 dB 58 percent of the time.

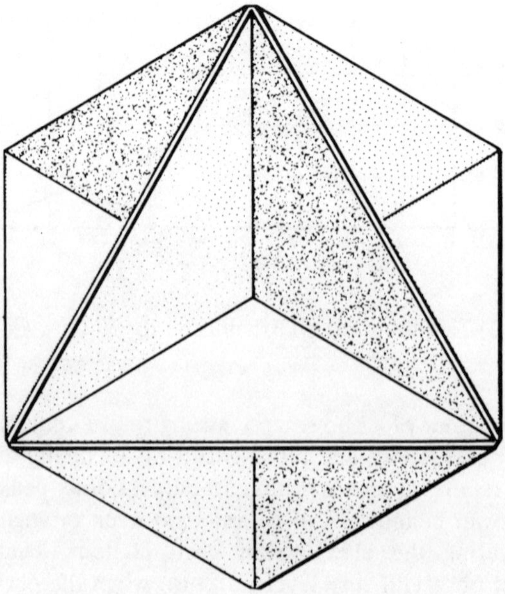

Figure 3–8. Cluster of Triangular-Corner Reflectors

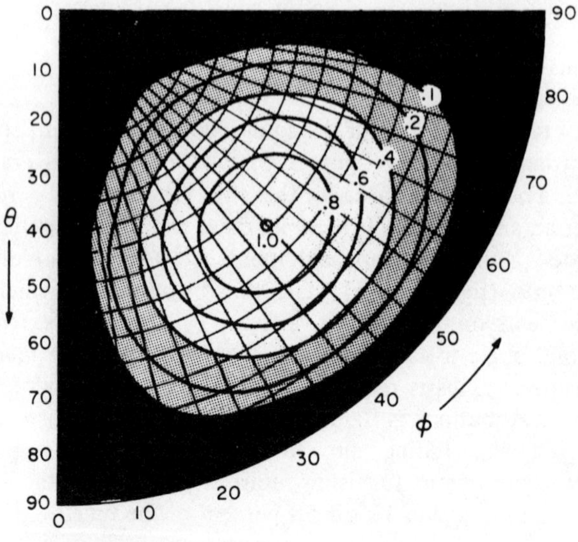

Figure 3–9. Triangular-Corner Return on Lambert Azimuthal Projection

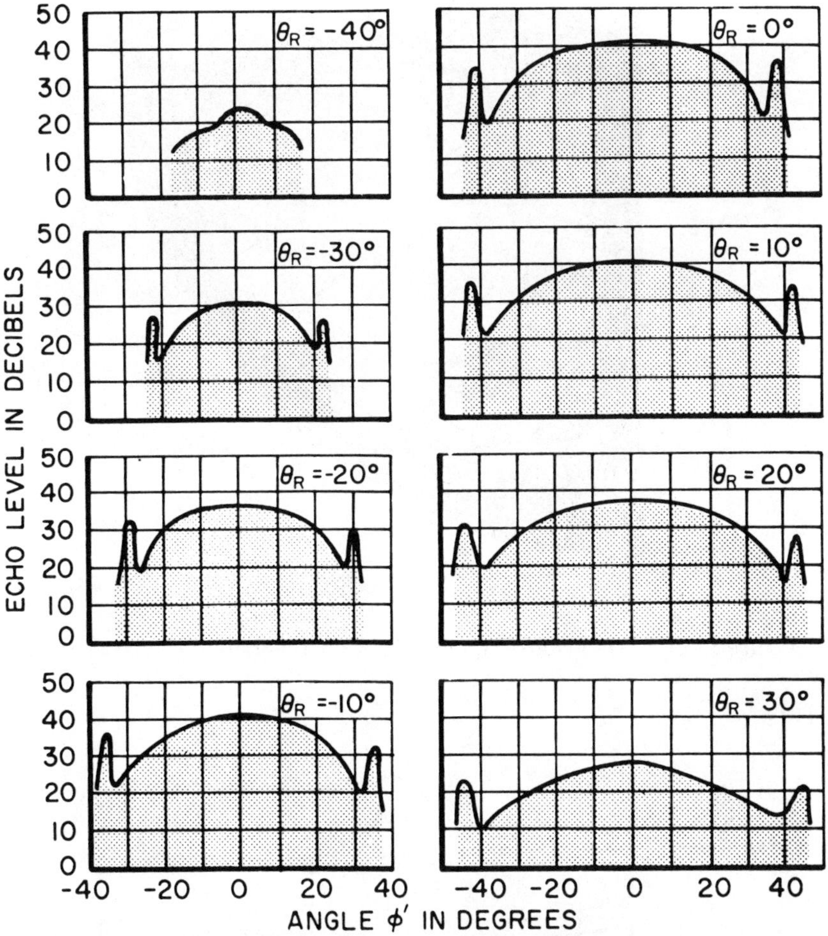

Figure 3–10. Measurement of Triangular-Corner Return

Circular-Corner Reflector

The circular-corner reflector (figure 3–13) has been analyzed by Siegel and coworkers at the University of Michigan.[3] The theoretical results are presented on the Lambert azimuthal grid in figure 3–14, where the return is in the form of a broad lobe with an angular width of 46° within 3 dB of the peak value. The probability distribution is given in figure 3–15, which also shows the distributions for the other corners. The return from the circular shape is superior to that of the other reflectors in its uniformity relative to the peak value. For random viewing, the return is with 3 dB of the peak 29

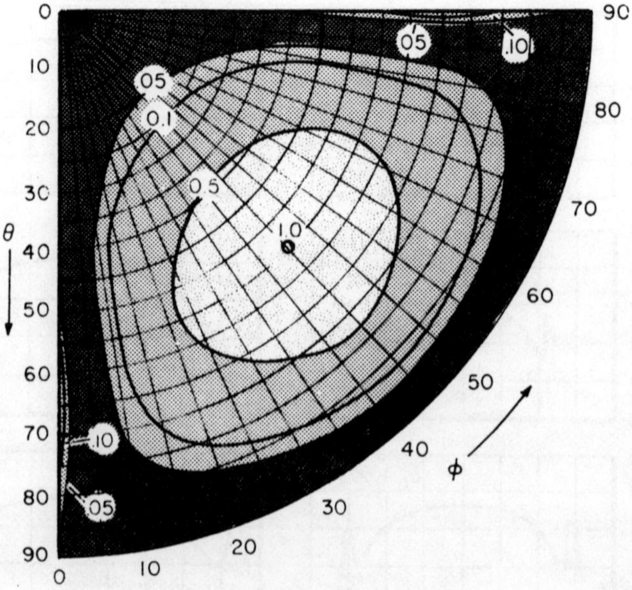

Figure 3–11. Triangular-Corner Return on Lambert Azimuthal Projection

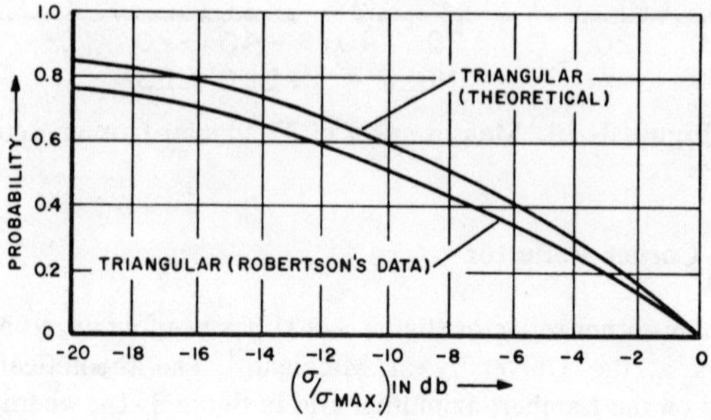

Figure 3–12. Cumulative Distribution for Triangular-Corner Return

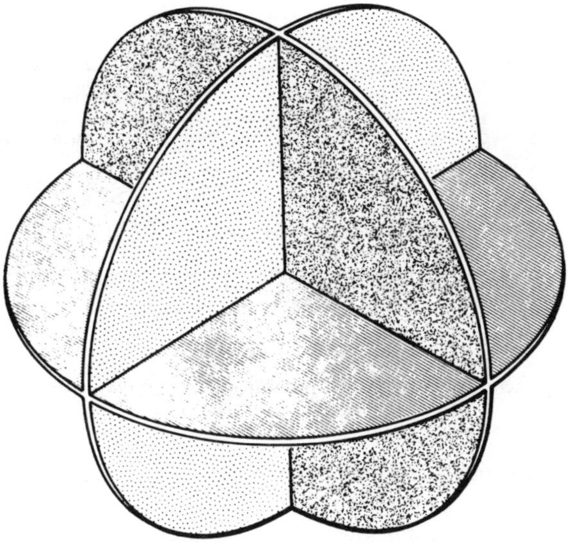

Figure 3–13. Cluster of Circular-Corner Reflectors

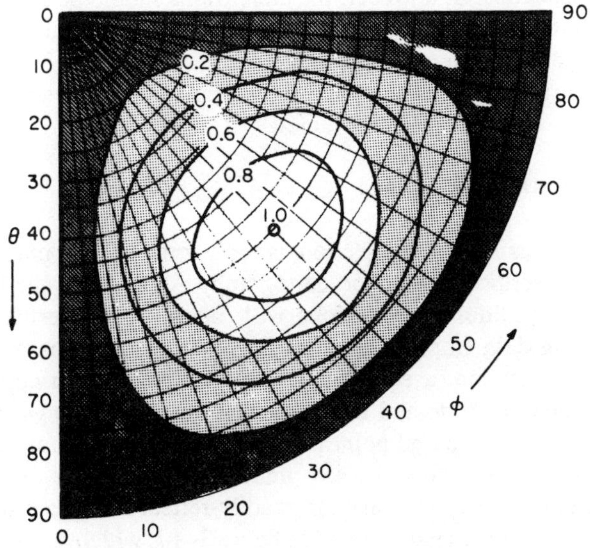

Figure 3–14. Circular-Corner Return on Lambert Azimuthal Projection

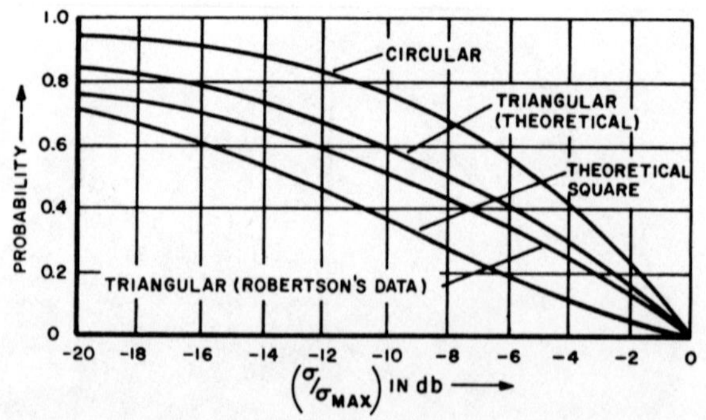

Figure 3–15. Cumulative Distribution for All Corner Reflectors

percent of the time, and the probability of exceeding −10 dB is 75 percent.

Figure 3–15 is valid for the physical-optics approximation, which requires that the dimensions be large compared with a wavelength. For this condition, the additional dihedrahl and flat-plate returns that predominate at certain angles do not extend over a great enough solid angle to influence the general conclusions. If, however, the dimensions should be of the order of the wavelength, then it is necessary to plot measured data in full. The evaluation of such data is further complicated by the variation of the return with polarization, a topic that is beyond the scope of this chapter.

Other Equal-Area Projections

The Lambert azimuthal projection is only one of the many equal-area projections described in texts on cartography A survey of the literature has revealed two additional projections with favorable properties for plotting backscattering data from bodies that have similar patterns in each octant. Collignon's projection, used in figure 3–16 of the circular-corner reflector, plots both the colatitude and the aximuthal circles as straight lines, so that interpolation is precise and point plotting is rapid. Furthermore, for objects that show symmetry about the 45° plane ($\phi = 45°$), after drawing up one-half of the contour pattern, one can readily reflect it through the 45° line.

The same data are again plotted in figure 3–17, which gives the Mercator equal-area projection. This grid has parallel, equally spaced colatitude lines; and the circles of constant azimuth are represented by sine cures. Both interpolation and reflection through the 45° azimuth are somewhat more difficult than with Collignon's projection, but the uniform spacing of the

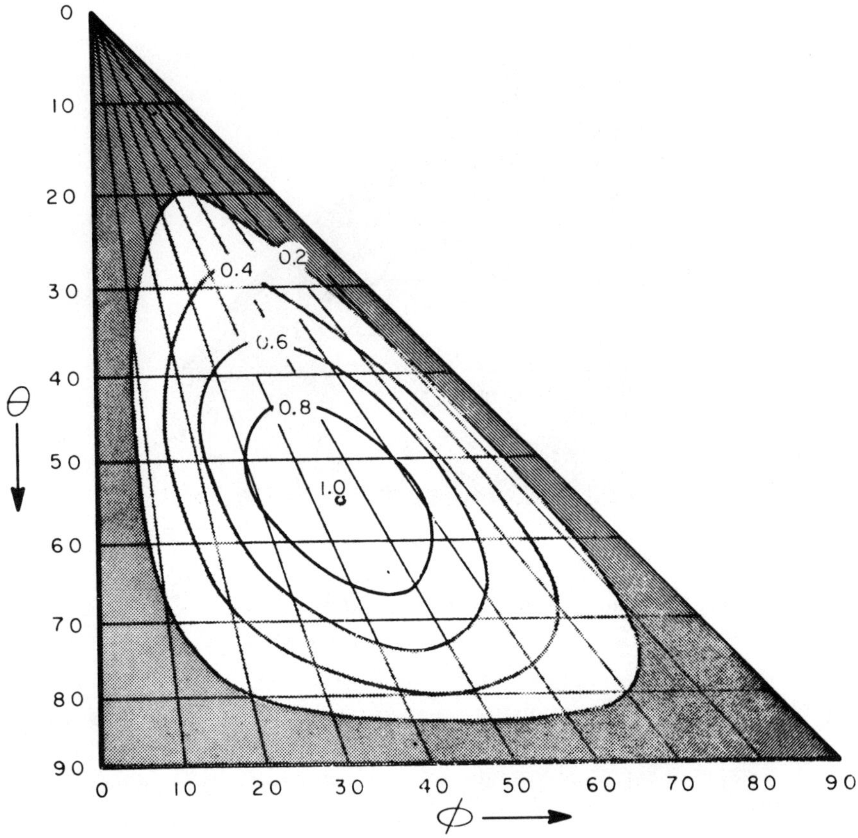

Figure 3–16. Circular-Corner Return on Collignon's Projection

angular intervals along the polar axis facilitates transfer of data from patter cuts that are run at constant ϕ angles.

On the subject of projections, note that the discussion here has been limited to contours that lie within one octant. For bodies having other reflective properties, such as a finite cone, or for the reflectivity pattern of a stablized satellite, the differing variation of the data suggests other grids as best suited for the presentation of contours [5].

A major criterion for selection of the projections has been convenience in plotting the data manually. Because of difficulty in plotting contours of more complicated reflective bodies, a computer program has been developed for this purpose. When the computer is used for automatic plotting, the emphasis in selection of a base grid is on ease of interpretation, since there is no

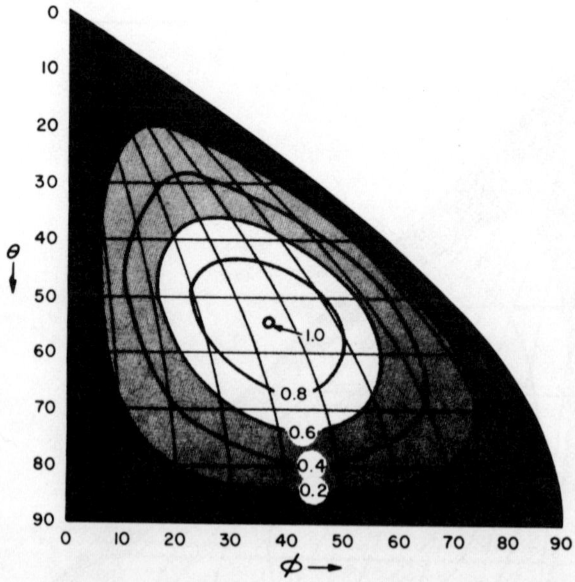

Figure 3–17. Circular-Corner Return on Mercator Sinusoidal Projection

problem of manual interpolation in plotting the points. In this case other projections may find favor. Figure 3–18 shows an early computer plot in the cylindrical equidistant coordinate system described in IRIG Standards [6].

Reflective Jacks

The corners have provided relatively simple variation of reflectivity with angle. It is of interest, therefore, to examine the handling of data for a passive reflector that has a much more detailed lobe structure, such as the jack illustrated in figure 3–19. Cuts were taken as several angles of tilt, resulting in a set of patterns similar to the one given in figure 3–20. These patterns have so much structure that it is impossible to visualize the spatial distribution of the backscattering lobes. The geometric relation of the plane of the cut to the (θ, ϕ) coordinate system is illustrated in figure 3–21. After the appropriate coordinate conversion, the data were plotted on Collignon's projection in figure 3–22, which clearly outlines the regions of stronger return. The cumulative probability is given in figure 3–23 and shows that the jack provides a very low level of return over most angles.

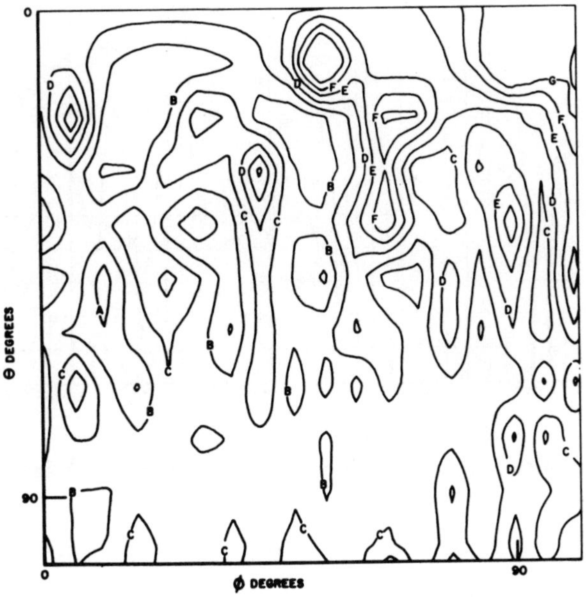

Figure 3–18. Computer-Generated Contours on Cylindrical Equidistant Projection

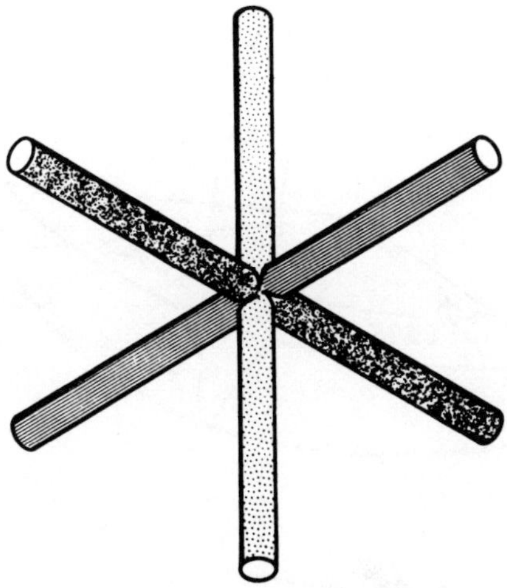

Figure 3–19. 5λ Jack

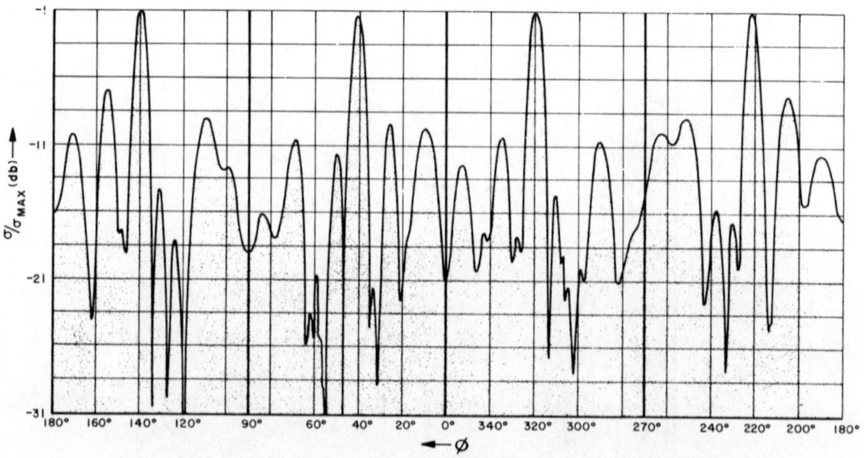

Figure 3–20. Typical Pattern Obtained for 5λ Jack

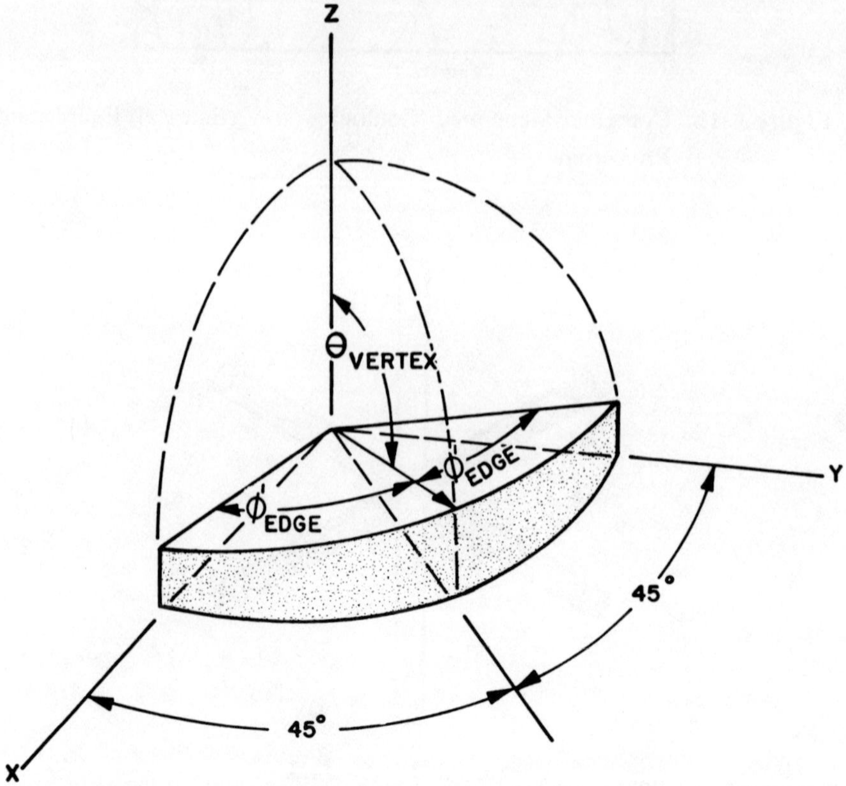

Figure 3–21. Geometric Relations for Great-Circle Cut

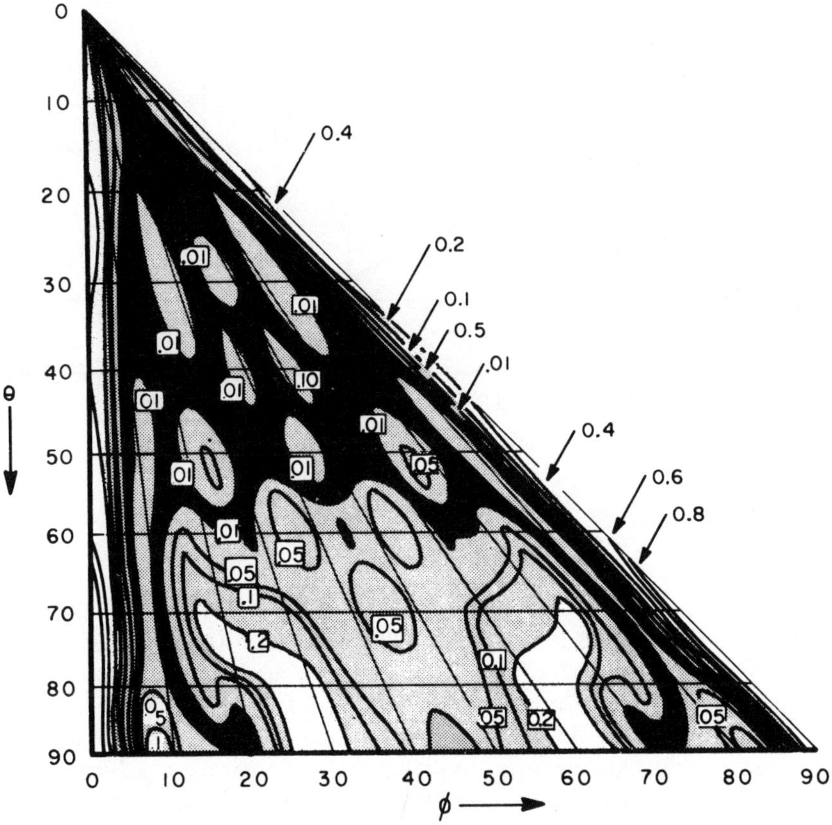

Figure 3–22. 5λ Jack Return on Collignon's Projection

Conclusions

By means of equal-area contour charts, the appropriate passive radar reflector can be selected for any specified angular coverage. For the case where the target is observed within the octant in which a simple corner is effective, Table 3–1 summarizes the characteristics of the three types of corners and the jack. The superior qualities of the circular corner would lead to its selection in most applications of corner reflectors.

In an attempt to increase the angular coverage from a passive reflector, one design evolved using a Luneberg lens. A series of spherical shells with relative dielectric constants chosen to follow a mathematical formula, which ensured focus of incident electromagnetic radiation at a point opposite the point of entry, was the basis of this design. A conducting partial hemi-

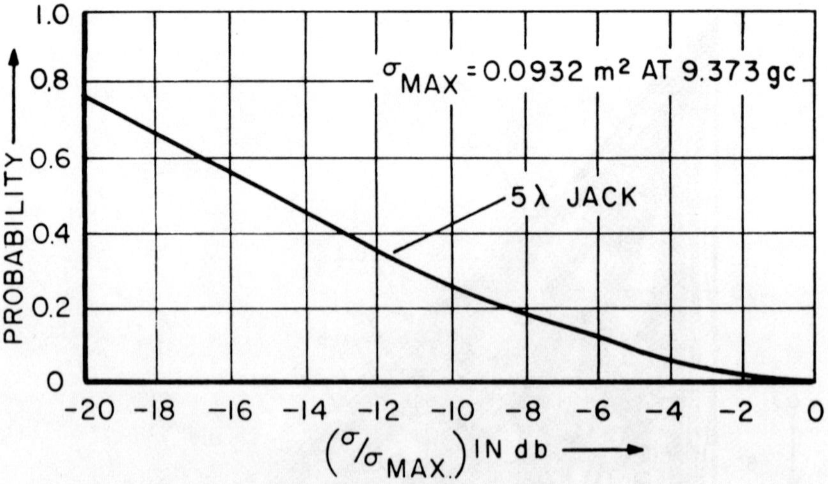

Figure 3–23. Cumulative Distribution for 5λ Jack

spherical shell then reflected energy back along the same path toward the source for wide angular coverage. The reflecting hemispherical reflector could be composed of a linear wire grid one-quarter wavelength from the solid conducting surface to return circular polarized fields in the correct sense to be received. In time it was discovered by experiment that it was not necessary that the lens have a varying dielectric constant: a homogeneous lens with a constant relative dielectric constant would suffice. Coverage of a typical design is shown here in figure 3A–1 through 3A–4. Dielectric rods with reflectors singly and in arrays also were used to provide specified angular reflective coverage.

References

1. Kennaugh, E.M. "Memorandum on Presentation of Radar Cross-Section Data by Contour Maps." Antenna Lab Report 475-8, Ohio State University, 30 October 1953.
2. Spencer, R.C. "Optical Theory of the Corner Reflector." MIT Radiation Laboratory Report R-433, 1944.
3. Siegel, K.M.; Barasch, M.L.; Crispin, J.W.; Goodrich, R.F.; Halpin, A.H.; Maffett, A.L.; Orthwein, W.C.; Schenstad, C.E.; and Titus, C.J. "Studies in Radar Cross-Sections XVIII—Airborne Passive Measures and Countermeasures." University of Michigan, January 1956, Appendix D.

4. Robertson, S.D. "Targets for Microwave Radar Navigation." *Bell System Technical Journal* 26 (1947): 852–869.
5. Levine, D. "Contour Charts for Radar Cross-Section Analyses." Symposium on Radar Reflectivity Measurement, MIT Lincoln Laboratory, 28–30 April 1964.
6. Inter-Range Instrumentation Group. "IRIG Standard Coordinate-System and Data Format for Antenna Patterns," IRIG Document no. 102–61, September 1961.

Appendix 3A: ECCO Reflector and ECCO Luneberg Lens

Ecco Reflector

The Ecco Reflector® is a passive reflector of radar energy (figure 3A–1). It has a large radar cross-section that is essentially constant over a wide conical viewing angle. The Ecco Reflector is compact, rugged, light in weight, and easily installed. It is broadbanded throughout the microwave frequency range.

The reflector is based on the Ecco Luneberg Lens. Energy incident on the lens is focused and reradiated in the direction from which it originated. In this respect it is similar to a corner reflector. The Ecco Reflector is far superior to the corner reflector for wide-angle coverage; it has a radar cross-section approximately eight times that of a circular-corner reflector of the same radius. Arrays of Ecco Reflectors produce scintillations and therefore can be used to simulate the reflectivity of objects of complex geometry.

The Ecco Reflector is available in almost any specified size in the range from 3″ to 48 ″ nominal diameter. It is supplied with a thin weatherproof radome.

Ecco Luneberg Lens: Three-Dimensional Microwave Lens

The Ecco Luneberg Lens (figure 3A–2) is a variable dielectric constant device of spherical shape that focuses an incident-plane electromagnetic wave to a point near its surface or, conversely, produces a plane wave from a point source. The Ecco Luneberg Lens has many unique applications. For example, rapid, wide-angle scanning of a radiation beam results from electronic switching of feeds over the surface of a stationary lens; many frequencies can be used simultaneously with the same lens. In general, beamwidth and sidelobe level of the Ecco Luneberg Lens, when used as an antenna, are approximately the theoretical values. Lenses are available in approximately the same sizes and weights as shown for the Ecco Reflector®. Each is enclosed in a thin, rugged, fiberglass-laminate radome. Lightweight lenses, available in some sizes, are approximately 1/4 the standard weight.

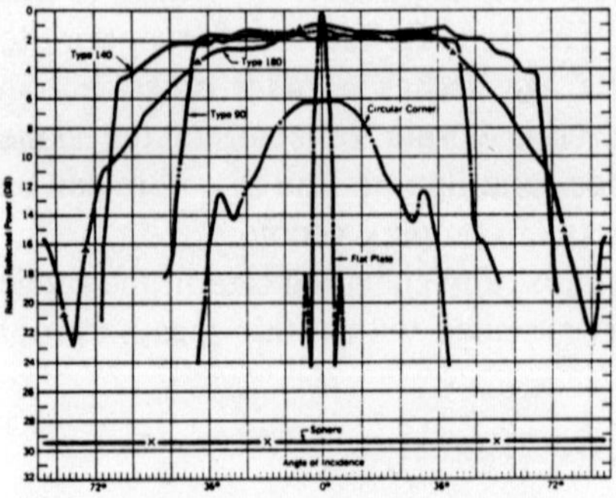

ANGULAR COVERAGE OF THREE TYPES OF ECCO REFLECTORS COMPARED WITH A CORNER REFLECTOR AND A METAL SPHERE. ALL ARE 12" IN DIAMETER. FREQUENCY IS 9.375 GHz (3.2 CM).

MODEL	LENS SIZE NOMINAL DIA. (INCHES)	THEORETICAL RADAR CROSS SECTION IN SQUARE METERS K BAND (1.5 CM)	X BAND (3.2 CM)	S BAND (10.0 CM)	WEIGHT LBS.
2B-103	3	1.16	0.255	0.026	0.3
2B-104	4	3.67	0.807	0.083	0.5
2B-105	5	8.96	1.97	0.202	0.9
2B-106	6	18.6	4.08	0.418	1.5
2B-107	7	34.4	7.57	0.775	2.3
2B-108	8	58.7	12.9	1.32	3.5
2B-109	9	94.1	20.7	2.12	4.9
2B-110	10	143.	31.5	3.23	6.6
2B-111	11	210.	46.1	4.72	8.5
2B-112	12	297.	65.3	6.69	10.9
2B-114	14	551.	121.	12.4	17.2
2B-116	16	940.	206.	21.1	25.5
2B-118	18	1505.	331.	33.9	36.4
2B-124	24	4758.	1045.	107.	86.
2B-136	36	24085.	5292.	542.	292.
2B-148	48	76121.	16726.	1713.	685.

VALUES ARE THEORETICAL. ACTUAL VALUES WILL BE BELOW THOSE SHOWN, PARTICULARLY FOR THE LARGER DIAMETERS AND HIGHER FREQUENCIES.

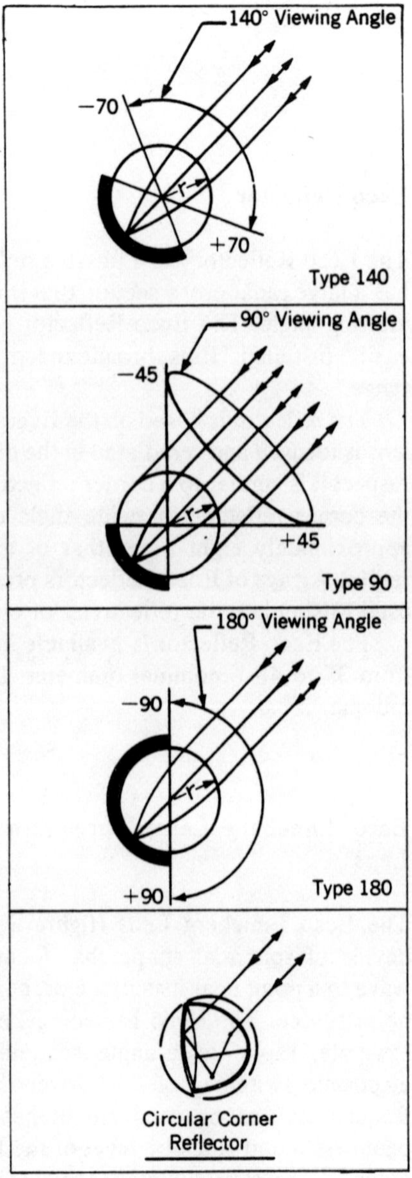

Figure 3A–1. ECCO Reflector Performance Data

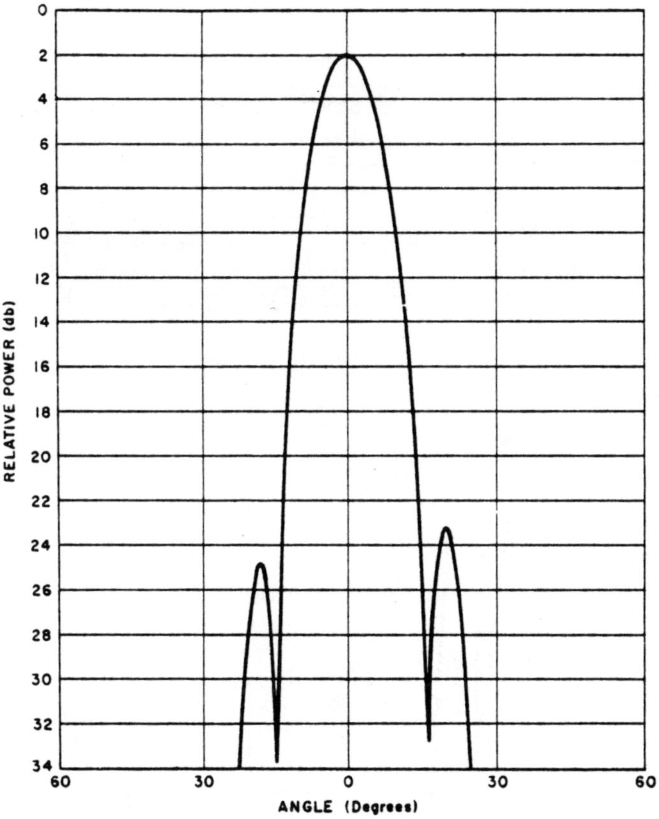

Figure 3A–2. ECCO Lens Antenna: H-Plane Pattern of 10″ Diameter Standard ECCO Luneberg Lens with Open-End Waveguide (WR 137) Feed at 6.0 GHz.

Bistatic Ecco Reflector

There are frequent requirements in aerial weapons testing for a device that will reflect microwave energy into a conical volume centered on the direction of the source. Such bistatic reflectors are essential whenever a radar system must function with an appreciable angular separation between the lines of sight connecting the target with the transmitter and with the receiver.

The Bistatic Ecco Reflector® is a spherically symmetrical dielectric lens similar to the standard or monostatic Ecco Reflector, with an associated spherical reflecting surface (figure 3A–3). Bistatic performance is obtained by a variation of the curve of dielectric constant versus radius. Bistatic Ecco

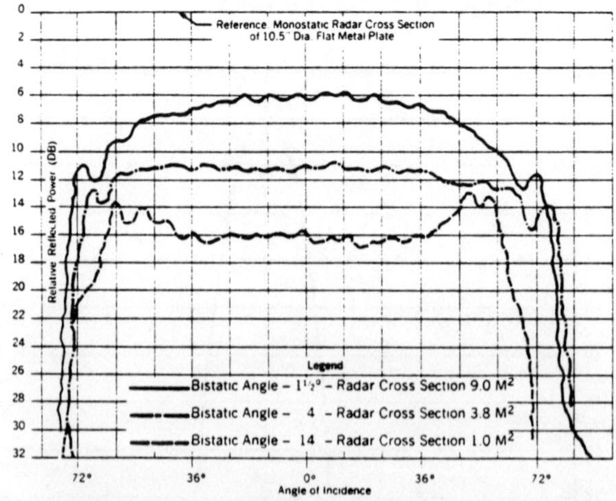

RADAR CROSS SECTION VS. ANGLE OF INCIDENCE FOR MODEL 2BS-210.5 (10.5" DIA.) TYPE 140 FOR SEVERAL BISTATIC ANGLES AT 9.375 GHz

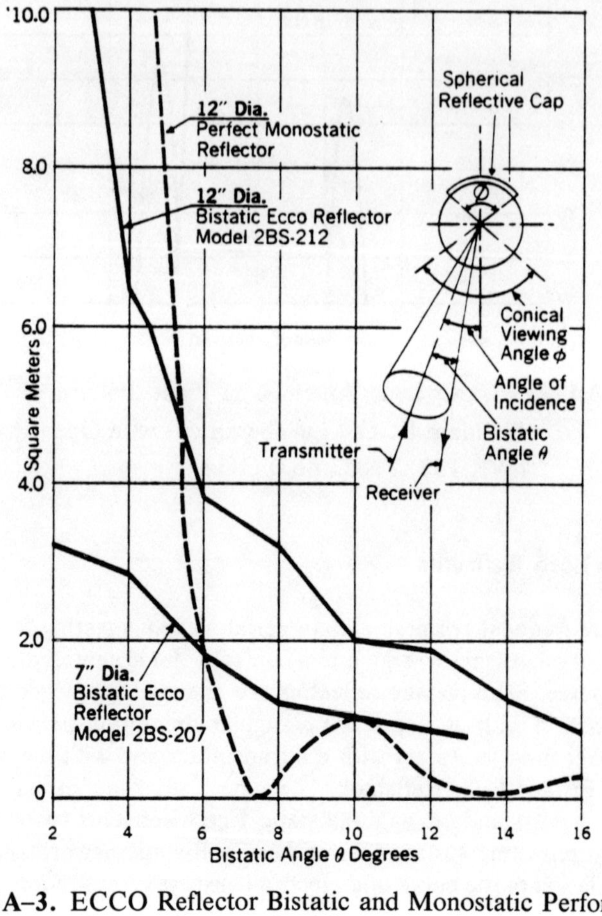

Figure 3A–3. ECCO Reflector Bistatic and Monostatic Performance

Reflectors are available from 3″ to 48″ nominal diameters, with various bistatic characteristics.

Omni-Azimuth Ecco Reflector

The Omni-Azimuth Ecco Reflector® is a modification of the Standard Ecco Reflector having a uniform radar cross-section for all values of azimuth angle and for elevation angles within the range approximately ±15° (figure 3A–4). It fills the long-standing requirement for a passive microwave reflector visible from all sides and having an elevation viewing range sufficient for use in seaway channel marking, in air-to-sea search and in small-aircraft cross-section augmentation for air-traffic control. Since the Omni-Azimuth Ecco Reflector exhibits complete uniformity of cross-section versus azimuth angle, clusters of reflectors are no longer required for complete azimuth coverage. Sizes are 3″ to 48″ nominal diameters.

		Typical Radar Cross-Section in Square Meters	
Model No.	*Diameter (Inches)*	*3.0 GHz*	*9.375 GHz*
2B-108-0A-30	8		4.0
2B-110-0A-30	10		8.0
2B-112-0A-30	12	1.2	18.0
2B-116-0A-30	16		35.0
2B-118-0A-30	18		55.0
2B-121-0A-30	21		80.0

Omni-Directional Ecco Reflector

The Omni-Directional Ecco Reflector® is a modification of the standard spherical Ecco Reflector that presents a uniform radar cross-section at all viewing angles. It is far superior to a cluster of corner reflectors in providing uniformity of response versus aspect angle. OD Ecco Reflectors are useful as buoy, channel or airfield markers, particularly where the required angular coverage is greater than that available with standard or Omni-Azimuth Ecco Reflectors. They can provide maximum wide-angle augmentation of the radar cross-section of small airborne vehicles. They are available with a socket mount in the form of a radial fiberglass tube bonded to the spherical surface; they can be suspended in an open-mesh rope net from any convenient tower or mast. Sizes are 4″ to 48″ nominal diameters.

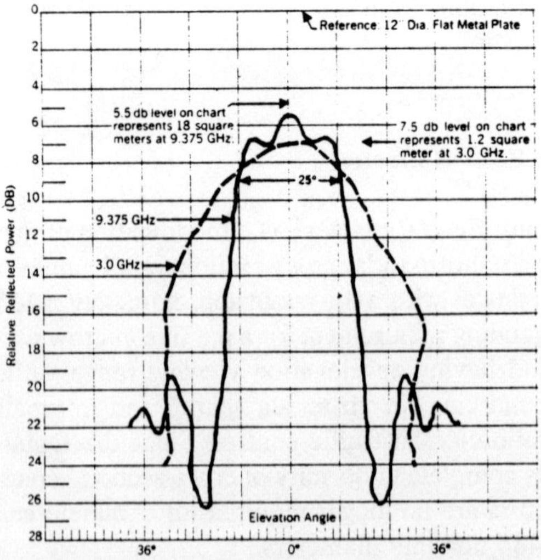

RADAR CROSS SECTION VS. ELEVATION ANGLE OF INCIDENCE FOR MODEL 2B-1112-OA-30 (12" DIA.)

Figure 3A–4. ECCO Reflector Radar Cross-Section

		Typical Radar Cross-Section in Square Meters	
Model No.	*Diameter (Inches)*	*3.0 GHz*	*9.375 GHz*
2B-104-0D	4		0.04
2B-105-0D	5		0.125
2B-106-0D	6	0.01	0.25
2B-107-0D	7		0.55
2B-108-0D	8		0.75
2B-109-0D	9		1.0
2B-110-0D	10		2.0
2B-112-0D	12	0.1	4.5
2B-115-0D	15		8.0
2B-116-0D	16		11.0
2B-118-0D	18	0.7	17.0

4 Radar Cross-Section Measurement Techniques

Introduction

In the past, measurement of low-cross-section bodies has been accomplished primarily by the use of precision indoor anechoic chambers, usually at high microwave frequencies. Use of shore distances between the antenna and the target, combined with either a doppler technique or a short-pulse technique, allows measurements [1] of cross sections of the order of $10^{-2}\lambda^2$, which is 10^{-5} m^2 at X-band. Measurements on indoor ranges [50] have been as sensitive as $10^{-4}\lambda^2$. The low-cross-section measurement requires an exceedingly careful cancellation of the background cross-section level, which is usually independent of the presence or absence of the physically small target.

For targets large in wavelengths, short distances cannot be used except through frequency scaling. However, scaling is undesirable for targets using lossy dielectrics, ablating materials, and radar-absorbing materials. In addition, the large size of the target tends to upset the background cancellation balance obtained in the absence of the target. When the cross-section of the target is commensurately large, the target-to-background cross-section ratio can be made suitably large. The combination of a large physical target and small cross-section—the goal for a variety of vehicles, including airplanes and missiles—presents a very difficult measurement problem because of the inherently high background level compared with the low target cross-section.

This chapter delineates the important problems incurred in realizing the aforementioned sensitivities for long distances, and lists five types of radar cross-section ranges. Each type is evaluated quantitatively in terms of the salient problems and is followed by a discussion of some typical equipment block diagrams.

Radar Cross-Section Ranges

Three salient problems can be extracted from the host of difficulties involved in measuring low radar cross-sections: (1) reducing background signal, (2) determining minimum distance between antenna and target, and (3) devising a support whose cross-section will not significantly affect the measurements.

These problems will be discussed in turn, followed by a discussion of each of the types of ranges and their potentialities.

Background Reduction

The extraneous signal that appears at the receiver terminals is called *background* and comes from many sources, mostly the result of nonspecular reflection of the transmitted energy into the receiving antenna. All the background energy results from the transmitted signal (receiver and antenna noise are relatively insignificant), whereby the sources of coupling include direct leakage or coupling in antenna/plumbing; specular reflections from ground or terrain; and nonspecular reflections from terrain, towers, platforms, cables, and so forth including diffraction from edges and bumps. Cross-section ranges fall roughly into five categories, depending on how the background is reduced: (1) anechoic chamber, (2) radome-enclosed range, (3) conventional outdoor range, (4) ground-plane range, and (5) oblique range. Each will be analyzed in turn after an examination of what background levels are tolerable and of what distances are needed.

The return in a radar cross-section range consists of the vector addition of the background field strength and the target field strength. The relative phase angle ϕ and the magnitude of both vectors change with the angular orientation of the target, often a limitation in the effectiveness of background cancellation schemes.

The following analysis demonstrates the vector addition of the target and background field strengths. Since the radar cross-section is proportional to the field strength squared, the total return is:

$$\bar{E} = \bar{E}_t + \bar{E}_b = \sqrt{\sigma_t} + \sqrt{\sigma_b}\, e^{j\phi} = \sqrt{\sigma}, \qquad (4.1)$$

where $\bar{E}$ = measured relative field strength,

$\bar{E}_t$ = target relative field strength,

$\bar{E}_b$ = background relative field strength,

σ_t = radar cross-section of the target,

σ_b = radar cross-section of the background,

σ = measured radar cross-section,

ϕ = phase angle between $\bar{E}_t$ and $\bar{E}_b$,

or

$$\sigma = \sigma_t + \sigma_b + 2\sqrt{\sigma_b \sigma_t} \cos \phi. \tag{4.2}$$

The limits of the measured cross-section are

$$\sigma_{\text{limits}} = \sigma_t + \sigma_b \pm 2\sqrt{\sigma_b \sigma_t}. \tag{4.3}$$

Thus a significant error can exist in the measured values when the target response is of the same order of magnitude or less than the background response.

Convenient approximate criteria for the relationship between background level and the target measurement accuracy are as follows: a measurement accuracy of ± 3 dB requires a background cross-section level 10 dB below the target cross-section; and a measurement accuracy of ± 1 dB requires a background level of 20 dB below the target cross-section. Figure 4–1 displays this relationship graphically, showing the upper and lower bounds of the errors in cross-section measurements.

The error curve of figure 4–1 is based on a nonvarying background level. The actual background level of an operating cross-section range can vary from day to day as a function of wind, season, humidity, wind, birds, insects, and other factors not readily amenable to tabulation.[1] Thus the phase angle ϕ is in general a function of time. Also, the phase angle can vary with angular orientation caused by the target and target-support interaction.)

Neglecting the short-term variations of ϕ, the target cross-section can be made independent of the phase angle ϕ by moving the target relative to the background. This can be illustrated by considering that the phase angle between $\bar{E}_t$ and $\bar{E}_b$ can be represented by

$$\phi = \frac{4\pi d}{\lambda},$$

where $d =$ path length difference (one-way path) between $\bar{E}_t$ and $\bar{E}_b$. Suppose the cross-section of the target is measured at two positions, d_1 and d_2; then the sum of the two readings at positions 1 and 2 is

$$\sigma_1 + \sigma_2 = 2\sigma_t + 2\sigma_b + 2\sqrt{\sigma_t \sigma_b}(\cos \phi_1 + \cos \phi_2). \tag{4.4}$$

To eliminate the dependence of the cross-section on ϕ_1 and ϕ_2, let

$$\cos \phi_1 = -\cos \phi_2.$$

$$\therefore \phi_2 = n\pi + \phi_1,$$

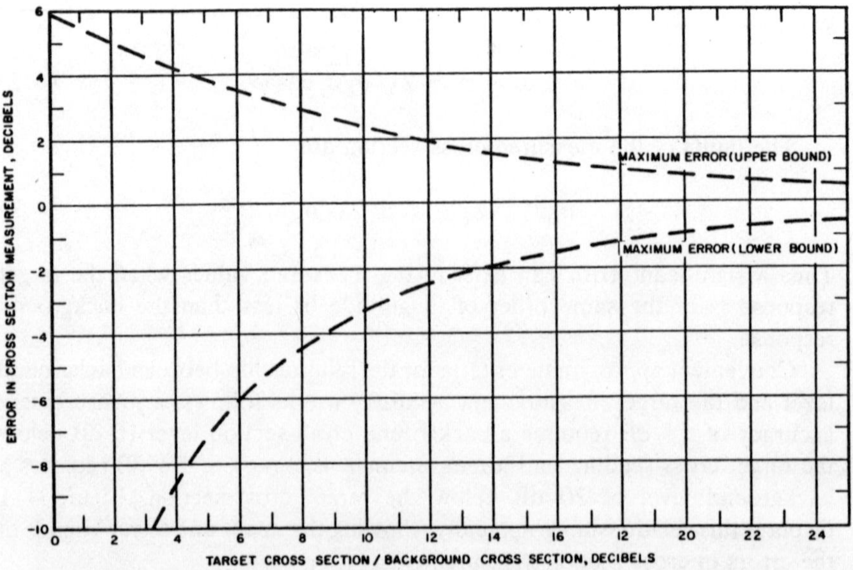

Figure 4–1. Measurement Error as a Function of the Ratio of Target Cross-Section to Background Cross-Section

where $n =$ some odd integer.

Expressing ϕ in terms of d,

$$\phi_2 = n\pi + \frac{4\pi d_1}{\lambda} = \frac{4\pi d_2}{\lambda}.$$

If the second position is at

$$d_2 = \frac{n\lambda}{4} + d_1,$$

then the target cross-section is

$$\sigma_t = \frac{\sigma_t + \sigma_2}{2} - \sigma_b. \tag{4.5}$$

Thus the target cross-section can be determined by measuring the cross-section of the target at two positions and with an independent background measurement.

This $\lambda/4$ technique has limitations because of the time dependency of ϕ as well as the angular orientation of the target. However, by repositioning the target to enable the radar return to pass from a maximum to a minimum signal, a ratio $\sigma_{max}/\sigma_{min}$ can be recorded. By referring to figure 4–1, the σ_t/σ_b corresponding to the measured $\sigma_{max}/\sigma_{min}$ can be determined. In this manner, a further check on the value of σ_t can be obtained.

To gain an appreciation of the order of magnitude of the background reflection coefficient required for a target of 10^{-3} m^2 (-30 dB m^2) or a background level of -49 dB m^2, for a ± 1-dB measuring accuracy, the following crude example is cited.[2] Assume that the background reflecting surface is within the entire antenna main beamwidth. If the antenna beam illuminates a flat metal plate of area A at normal incidence, the scattering cross-section is $\sigma = 4\pi A^2/\lambda^2$. For a target size of 0.5 m $\times$ 0.5 m, the plate at L-band ($\lambda = 0.3$ m) would be roughtly 8.8 m^2 ($\simeq$ 9.4 dB m^2). Thus, to achieve the background level of -49 dBm^2, a net reflection coefficient of -58.4 dB is required. This net reflection coefficient can be the resultant of multiple or single bounces of the reflected wave or of any other type of cancellation techniques. The foregoing calculations assumed background return only from main beam illumination. In addition, sidelobe illumination must be considered, as these lobes can illuminate highly reflecting objects.

Distance

The minimum distance between the antenna and the target is one of the important parameters in the choice of radar cross-section measuring ranges. From a measurement point of view, short distances require low power and reduced background problems. In a typical actual situation involving a radar looking at an aircraft or missile, the target is used in the far-field region, defined as that region in which the antenna angular pattern is substantially independent of distance. For both array and quasi-optical antennas (such as a parabolic dish) a simple criterion serves to demarcate the far-field region. For a uniformly illuminated aperture of diameter D, the aperture looks like a point source to within an error of 22.5° ($\lambda/16$) at a distance given by $R = 2D^2/\lambda$. That is, the phase front produced by the antenna at this distance is constant within the 22.5° error over a region of the same size as the aperture. The antenna directivity (often called *gain*) at this distance is within 1 percent of the value at an infinite distance. At distance D^2/λ, the maximum phase error is 45° and the directivity is 6 percent low [2]. For an aperture with tapered illumination, which represents almost all practical cases due to the greater sidelobe ratio that can be obtained, the effective aperture is decreased over that for uniform illumination. Similarly, the distance at which a 22.5° phase error occurs is also decreased. For a circular aperture with taper of $(1 - \rho^2)$, the effective aperture is 0.75 times the actual aperture, and the far-field criterion is now $R = \pi D^2/2\lambda \simeq 1.5\ D^2/\lambda$.

The distance $2D^2/\lambda$ is usually taken as the start of the far-field region; everything within is the near-field region. As the observer progresses from the far region into the radiating near-field region, two effects are noted.[3] First, the main beam broadens. Second, the nulls between lobes fill in and the sidelobes are raised slightly [3]. These effects become progressively worse as the distance decreases, until the point at which the main beam tends to bifurcate into two beams, depending on the aperture and aperture distribution, in a manner analogous to the overcoupled circuit [4, 5, 13].

However, antennas can be utilized within the radiating near-field region by a technique known as *focusing* [6]. Here the antenna is designed so that the rays focus at a finite point in space (conventional antennas are focused at infinity). Focusing produces in a region about the focal point the exact pattern normally found in the far field. The pattern exhibits a depth-of-field behavior analogous to that of optical lenses where the pattern deteriorates as the observer moves closer to the aperture or farther away from the aperture. Two distances often referred to are the ***Rayleigh distance*** [7] of D^2/λ and the *hyperfocal distance* of $D^2/2\lambda$. Antennas have been used successfully as close as $0.03\ D^2/\lambda$ [8]. One advantage of focusing pointed out by Rayleigh is that a more constant phase front can be obtained over the focal region [9].

Two distinct steps in the antenna-target scattering process can now be identified. First, the antenna illuminates the target; second, the antenna receives the energy scattered by the target. Criteria will be developed for both these steps. In the first case it is necessary that the target be immersed in a field sufficiently like a plane wave so as to simulate accurately the long-range radar situation. Figure 4–2 shows the amplitude variation over a target as a function of the ratio of target to antenna diameter [10]. In figure 4–3 the distance is depicted as a function of aperture size for 2.0, 1.5, and 1.0 D^2/λ at $\lambda = 1$ ft (L-band, roughly 1 GHz). These give respectively maximum phase variations over the aperture of 22.5°, 30°, and 45°. The 1 Gc frequency is typical; the data can readily be scaled to other frequencies. The choice of values among these curves will be discussed in detail. In addition to the transverse-amplitude variation, there is a longitudinal change in amplitude because the target length is not negligible compared to the $1/R$ variation. Figure 4–4 shows the amplitude variations as a function of range for several sizes of targets. For distances that satisfy the transverse criteria, the longitudinal-amplitude variation is very small.

For purposes of analysis, the scatterer may be thought of as a new antenna whose radiating-current distribution is produced by the incident wave. In order for the scattered energy to exhibit characteristics analogous to those of the antenna field in the far region, the receiving antenna must be a certain minimum distance from the target. For a spherical scatterer, the

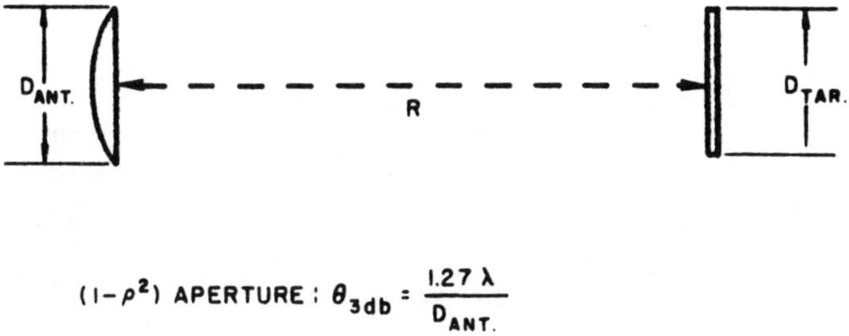

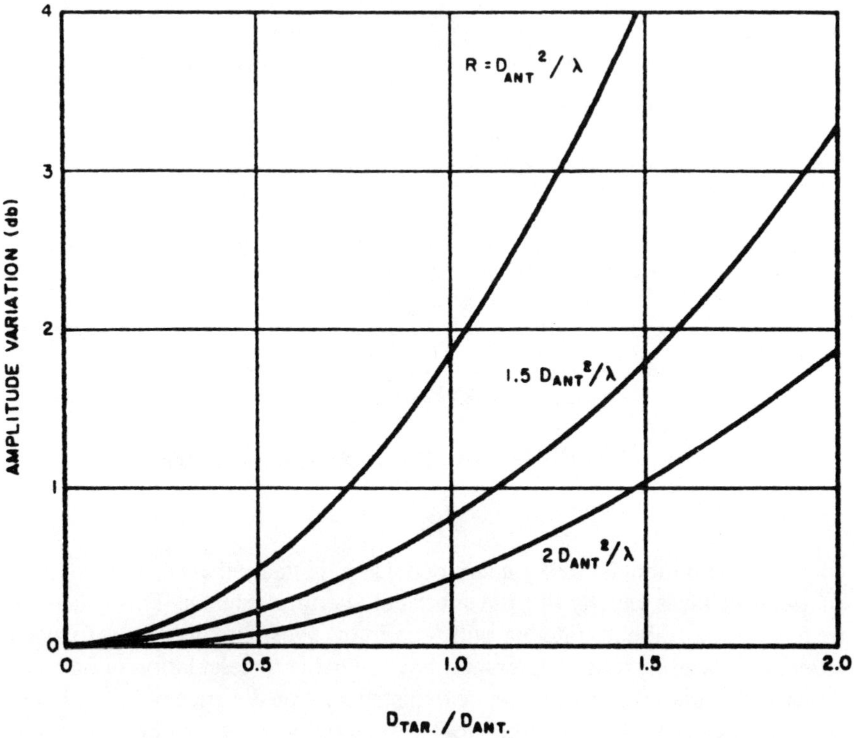

Figure 4–2. Transverse Amplitude Variation over Target

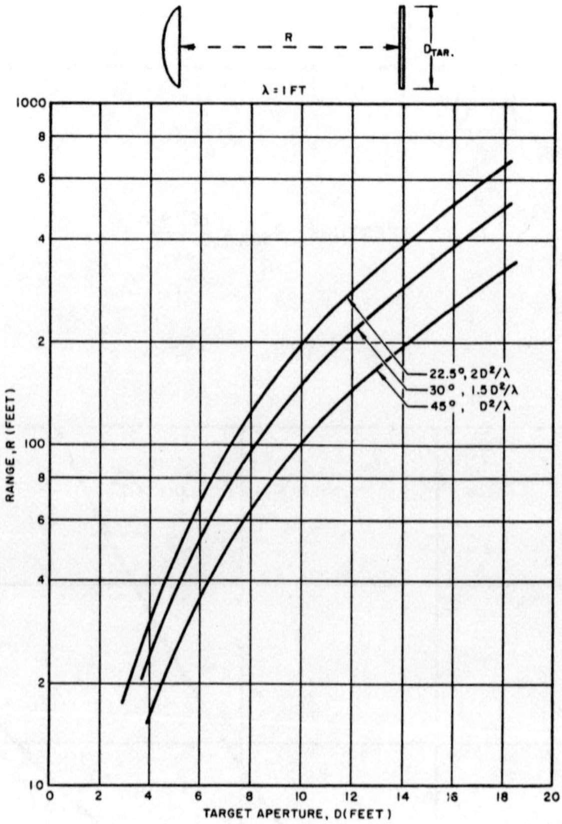

Figure 4–3. Transverse Phase Variation over Target

current distribution is well understood [11]; in fact, the far field of this scatterer behaves exactly like the spherical-aperture antenna. This is usually treated as a parabolic antenna with quadratic phase error or as a focused antenna. However, most scatters are not amenable to calculation of how the pattern of scattered energy varies with distance. At a distance of $2D_{tar}^2/\lambda$, the wave scattered from the target looks essentially as it would an infinite distance away. This wave must be received by the antenna; and, for the antenna to operate in the expected manner on an incident plane wave, the plane wave must, as noted in th previous discussion on circular apertures, be $1.5D_{ant}^2/\lambda$ away. Thus the range between antenna and scatterer can be thought of as consisting of two distances added together. So the total range is given by

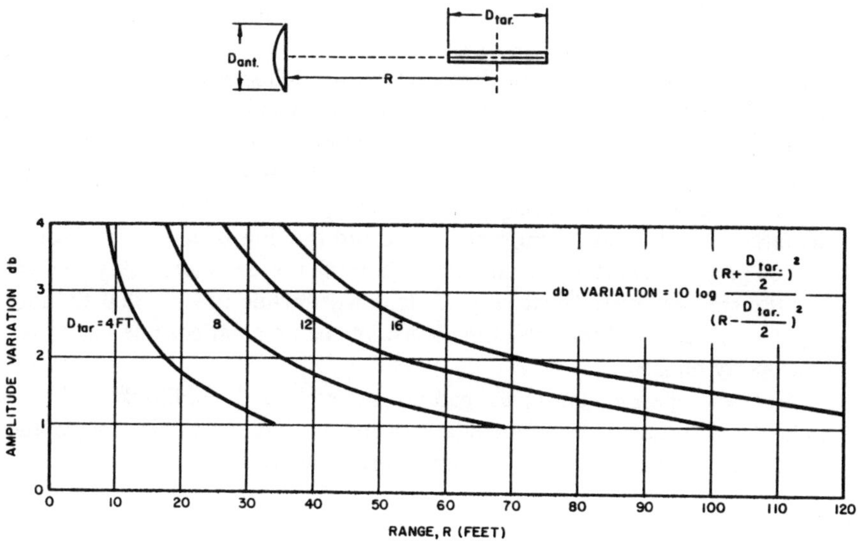

Figure 4–4. Longitudinal Amplitude Variation over Target

$$R = \frac{1.5D_{\text{ant}}^2}{\lambda} + \frac{2D_{\text{tar}}^2}{\lambda} \tag{4.6}$$

for an unfocused antenna. As mentioned earlier, use of focused antennas would allow the first term to be somewhat reduced.

If the antenna term in equation 4.6 is reduced, the receiving properties will not be seriously disturbed until the sidelobe enhancement causes pick-up of extraneous signals. This will occur long before the gain reduction due to main-beam broadening is serious. Thus, for receiving purposes; the antenna need not be focused and can be operated fairly close. However, it must be remembered that the total distance consists of the term involving the antenna plus the term involving the target, and the latter cannot generally be reduced.

From figure 4–3, a target of 10 ft in length requires a minimum range of approximately 350 ft at L-band with a conventional 10-ft diameter antenna. Even with focused antennas, about 250 feet of range would be needed. At C-band of 10-ft target would require about 1,700 feet of distance.

More generally, the range criterion can be expressed as

$$R \geq pD_{\text{tar}}^2/\lambda, \tag{4.7}$$

where it is assumed that D_{ant} and D_{tar} are equal or nearly so [12]. D_{tar} is the maximum projected transverse dimension, for the monostatic case. The dimensionless quantity, p, has been empirically chosen to be between 0.25 and 4. Use of $p \simeq 0.5$ has allowed radar cross-section data acquisition not too noticeably different from that obtained when larger values of p have been used [14, 51]. The smaller values of p will compromise the fine structure in the resulting cross-section patterns, depending on the size and form factor of the target. If it is deemed important to record the fine structure of the cross-section pattern versus aspect angle faithfully, the price in terms of real estate and transmitter power must be paid. It is worthwhile to note that turntable angular velocity and recorder time constants can also affect fine structure if improperly chosen.

Recent experimental investigations by Kell and Hendrick [51] have yielded the following recommendations for various target sizes if detailed fine-structure characteristics are not of prime interest.

D_{tar}/λ	R
3	$4\,D_{tar}^2/\lambda$
10	$1/2\,D_{tar}^2/\lambda$
50	$1/4\,D_{tar}^2/\lambda$

Target Support

The conventional method of supporting targets uses a turntable on which is mounted either a polyfoam (or other dielectric) tower, or a steel column surrounded by suitably shaped absorbing material. This assembly is sometimes located on an earth mound or wooden platform in order to increase target height above the ground and thus minimize ground-reflection effects. Platforms and mounds are undesirable for the low-cross-section measurement problem because of scattering and diffraction by the structure. In many cases it is desirable to mount the turntable below ground level and to have the turntable top flush with the surface.

Both the styrofoam and the best absorbers have a plane-wave reflection coefficient at normal incidence of about −35 dB.[4] Tapered styrofoam column return can be reduced by approximately 20 dB over that from a straight styrofoam column [15]. At L-band a column of absorber is typically several feet in diameter, and a foam column is of comparable size. Styrofoam columns can be used to support intermittent loads of about 1,000 pounds. Since these columns can be expected to be at the best only 55 dB below the same structure of metal, such foam or absorber-covered columns are therefore marginal.

A more attractive scheme, using nylon rope to support the target, has been utilized with success by the Royal Aircraft Establishment (RAE) [16]. However, even nylon rope has an appreciable radar cross-section. Measurements on nylon supports have been made by the Cornell Aeronautical Lab (CAL) at 9.2 GHz using nylon rods [14] and by Cram et al. at 35 GHz using nylon string [17]. In both cases several diameters were measured at a single frequency for normal incidence with both parallel and perpendicular polarization. The measurement technique involves the illumination of the string by an antenna; and, in order to calculate the cross-section per unit wavelength (so that a generalized formula may be found), it is necessary to compensate for the nonuniform illumination in both amplitude and phase of the string. To a good approximation the illuminated length, L, can be taken as that between the 3-dB points on the antenna pattern; the string outside this region contributes a small backscatter that is nearly equal to that lost because of nonuniform illumination in the beamwidth. Since a rope is a line scatterer, the cross-section should vary as L^2, with the exception that defocusing may have to be considered (see equation 4.10). A normalized parameter of interest is σ/λ^2 per unit wavelength squared of rope and may be expressed as

$$\frac{\sigma}{\lambda^2(L/\lambda)^2} = \frac{\sigma}{L^2}. \tag{4.8}$$

The data from CAL and from Cram have been reduced to this normalized form and plotted (see figure 4–5). In the CAL case, using the beamwidth and distance given, the result is

$$\frac{\sigma}{L^2} = 60\left(\frac{d}{\lambda}\right)^{3.4}, \tag{4.9}$$

where d is the diameter of the rope.

In the Cram et al. measurements, a distance of 45 ft was used.[5] With the beamwidth given, this yields a value of $L/\lambda = 320$, and this value was used in reducing the data. When the two values were first compared, there was a great dispaity, apparently caused by defocusing of the energy scattered from the rope. Although the rope is in the far field of the illuminating antenna, the receiving antenna (the same antenna, of course) is in the radiating near field of the line scatterer [18]. Thus the dependence $\sigma \sim L^2$ must be diminished by the defocusing factor, which for this case is essentially the asymptotic value of

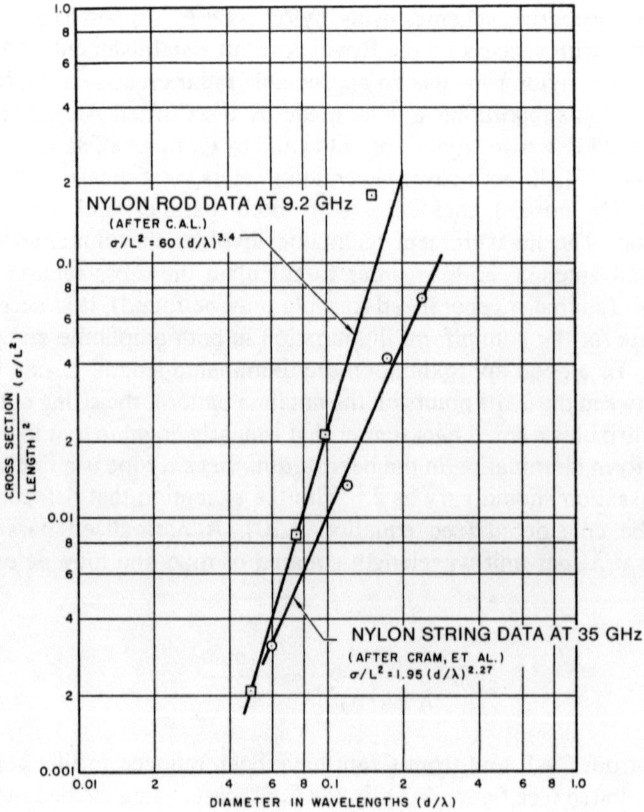

Figure 4–5. Comparison of Dielectric Cylinder Radar Cross-Section Data, for Parallel Polarization and Normal Incidence

$$\text{Defocusing Factor} = \frac{\pi L^2}{8\lambda R}. \tag{4.10}$$

This then results in

$$\frac{\sigma}{L^2} = 1.95\left(\frac{d}{\lambda}\right)^{2.27}, \tag{4.11}$$

and it can be seen that the agreement, although not good, is encouraging. In addition, a braided rope has a slightly lower effective dielectric constant compared with solid material. Thus it appears that scattering cross-section of nylon ropes can be roughly scaled from one frequency to another.[7] However, more experimental data are desirable. The importance of avoiding parallel polarization and normal incidence is clear. Tape might offer advantages if the orientation were not normal to the incident energy.

It is interesting to compare the effect of support rope at different frequencies. For frequency scaling the effects are independent of frequency, except for high-loss ropes. For full-scale measurements at different frequencies, the target weight, minimum rope strength W, and size d are fixed. The range distance is proportional to λ^{-1} from equation 4.6; and the length of rope L will be $\sim R$ for fixed antenna beamwidths. Thus $L \sim \lambda^{-1}$. Since rope strength $W \sim d^2$,

$$\sigma \sim L^2 \frac{d^{3.4}}{\lambda^{3.4}} \sim \frac{W^{1.7}}{\lambda^{5.4}} \sim \lambda^{-5.4}, \qquad (4.12)$$

so that the full-scale support problem becomes markedly easier at lower frequencies. For a numerical example, CAL quotes a diameter of 0.071 cm for a 50-lb nylon line [14]. A 50-ft length at 1 GHz of 500-lb line would have total cross-section of $\simeq 10^{-3}$ m^2.

Equation 4.12 may be evaluated using a known relationship as just given between nylon monofilament cross-section and tensile strength to yield estimates of residual background cross-section as a function of fractional weight supported with wavelength as a parameter. The resulting empirical formula is

$$\frac{\sigma}{\lambda^2} \simeq 6(10^{-14}) \left[\frac{\sqrt{W}}{\lambda} \right]^{3.4} \qquad (4.13)$$

where W = tensile strength in pounds, and

λ = wavelength in centimeters.

The curves resulting from equation 4.13 are shown in figure 4–6. It must be emphasized that these data constitute worst-case estimates, and the results do not consider other form factors for nylon or use of polarization and control of angle of incident illumination to minimize reflections.

In operation, the target and supporting lines may be wound up slightly so that when released the target rotates slowly in azimuth. If desired, precise angular information can be obtained using motion picture of an infrared (IR) technique. The latter would use an index mark of high emissivity on the target, with an IR scanner. Normally, precise angular position measurement is of no value. For almost all cross-section measurements, including correlation with theoretical results, a 1° accuracy appears more than adequate. In fact, a range sufficiently good to measure low cross-sections

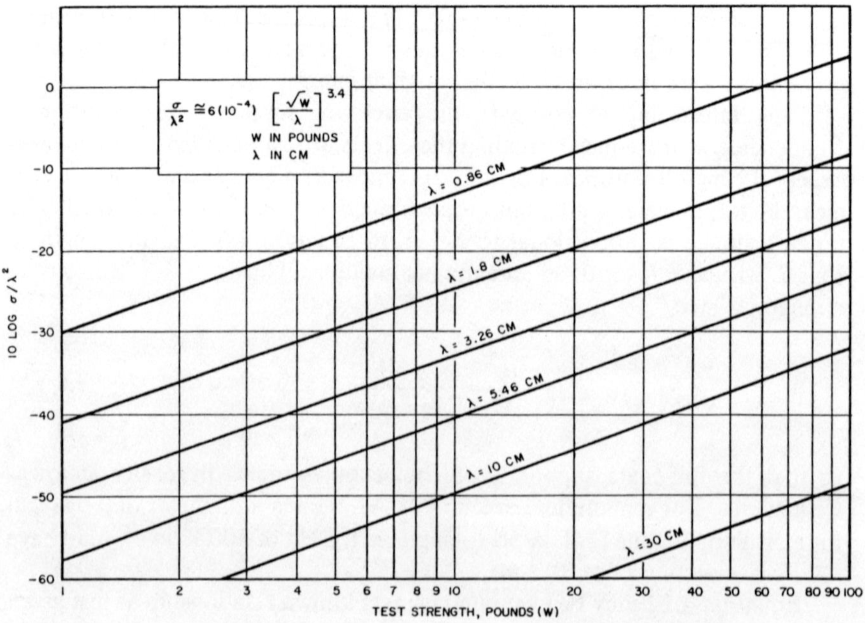

Figure 4–6. Estimated Cross-Section of Nylon Line as Function of Test Strength and Radar Frequency

such a pattern can easily be located through symmetry. Thus the rope support system is more than adequate in target positioning. Rope or a styrofoam tower in high winds may have some effects on the measuring accuracy. It must be emphasized that any large-aperture, low-cross-section target support will probably have a wind problem. However, the wind problem can be minimized by a large, air-inflated radome that surrounds the support and/or the rope support towers.

Regarding heavy targets, the requirements of low cross-section seem to be in conflict with the properties of a support. It seems exceedingly difficult, if not impossible, to measure low-cross-section values on heavy targets except during actual flights. Rope supports, however, have been used satisfactorily for a few hundred pounds and can probably support much more. The ropes could be attached through holes that are small in diameter compared with the wavelength. The support towers must be outside both the range gate and the antenna main beam, and must not contribute to multipath reradiation.

Ideally, the target cross-section should be measured without any supports. The cross-section of in-flight aircraft has been measured successfully over limited aspect angles (radar line of sight) as a function of range [41]. However, such methods are feasible only for specialized purposes.

Exotic schemes such as the use of supporting air columns have been demonstrated to be technically feasible for small targets and analysis has indicated feasibility for much heavier targets [42]. Such schemes might have initial application in supporting extremely small cross-section calibration targets to estimate target-support contribution to total background cross-section.

RCS Measurement Considerations

Anechoic Chamber

The anechoic chamber is simply a room lined with absorbing material. The ordinary rectangular parallelepiped chamber admits, in addition to the direct ray (from scatterer to antenna), a number of single-bounce rays. An improvement on this design is realized by making walls in the shape of a shallow wedge or pyramid [1]. The late Dr. C.B. Barker showed that such a room can be designed using ray optics so that with a point source and a point sink, all single-bounce paths can be eliminated. Thus the background is attenuated twice by the absorbing material. Unfortunately, finite-size apertures, edge diffraction, and the like degrade this optimum performance somewhat. An alternative scheme uses either transverse or longitudinal absorbing baffles [37]. The longitudinal baffles can also be arranged to eliminate all single-bounce paths in an anechoic chamber.

To obtain an idea of the quality of such rooms, figure 4–7 shows data for a typical high-grade absorber. From these data it can be seen that at least 12 dB of attenuation is maintained over incidence angles from normal to 75°, thereby giving a 24-dB attenuation for the ideal double-bounce chamber up to 75° from normal incidence. This 24 dB, together with an antenna sidelobe level of −25 dB, gives −49 dB of signal attenuation for two side bounces. Thus the residual background is sufficiently low to measure a target of 10^{-3} m^2 cross-section with a ±1 dB accuracy. Figure 4–8 shows room width as a function of length for a minimum incidence angle of 15°. The length and width required and the background level of the anechoic chamber may preclude the use of an indoor range for large targets because of economical considerations.

Two instrumentation techniques often used in anechoic chambers will be described in detail later. The background-cancellation scheme uses a CW radar, with the background signal canceled out by use of a hybrid junction, short-slot coupler, and so forth. The cancellation signal must have the proper

EXTRAPOLATED TO L-BAND, 26" THICKNESS USING B.F. GOODRICH
DATA FOR 7500 MHz, 3 1/4" THICK., TYPE VHP

CURVE REPRESENTS THE LOCUS OF MINIMUM AMPLITUDE OF SEVERAL
CURVES SUPPLIED BY B.F. GOODRICH.

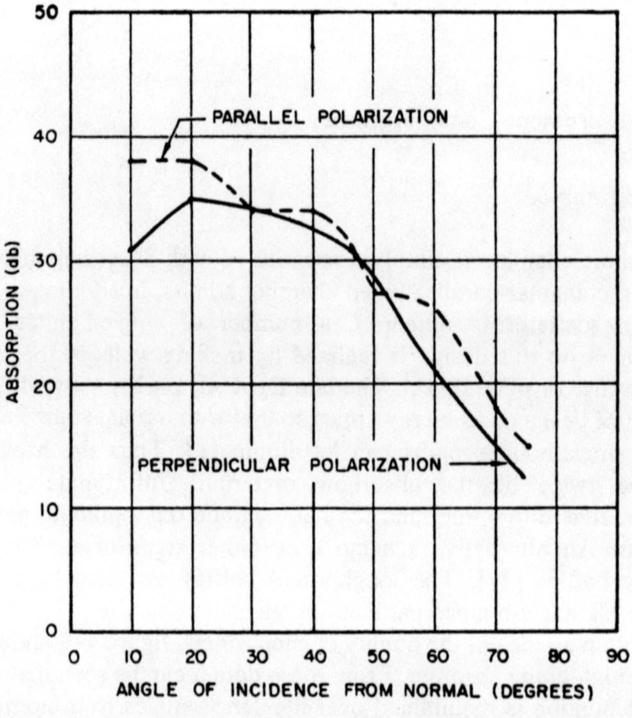

Figure 4–7. Absorption Characteristics of Microwave Absorber

phase and is usually at the carrier frequency. The second technique involves oscillating or nutating the target to impart a doppler component to the signal. Selective filtering of this audio frequency component allows suppression of the nonmodulated background. The doppler method requires a more uniform electromagnetic field and may be difficult for bodies of large inertia.

Radome- or Balloon-Covered Range

Radomes or large air-supported plastic structures can provide climatic control and protection from other elements of nature that might affect the

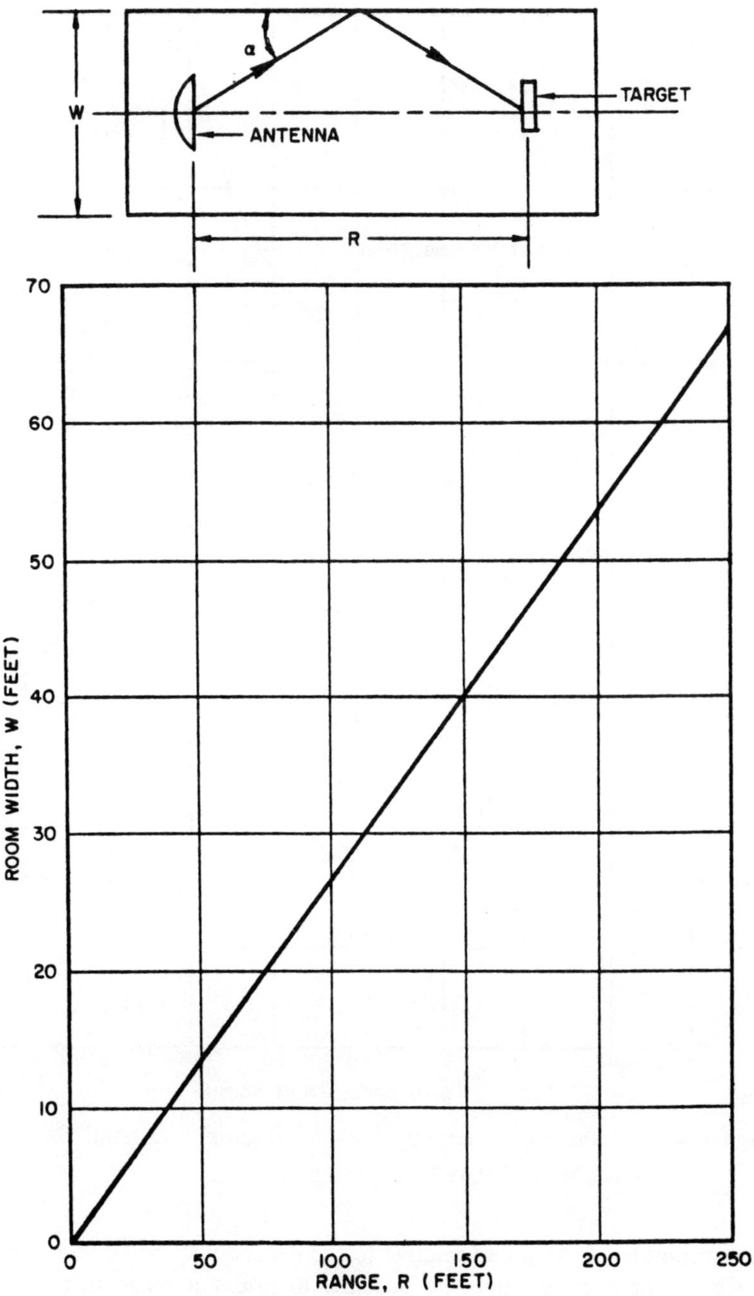

Figure 4–8. Minimum Room Width versus Length for Grazing Angle, $\alpha \geq 15^\circ$

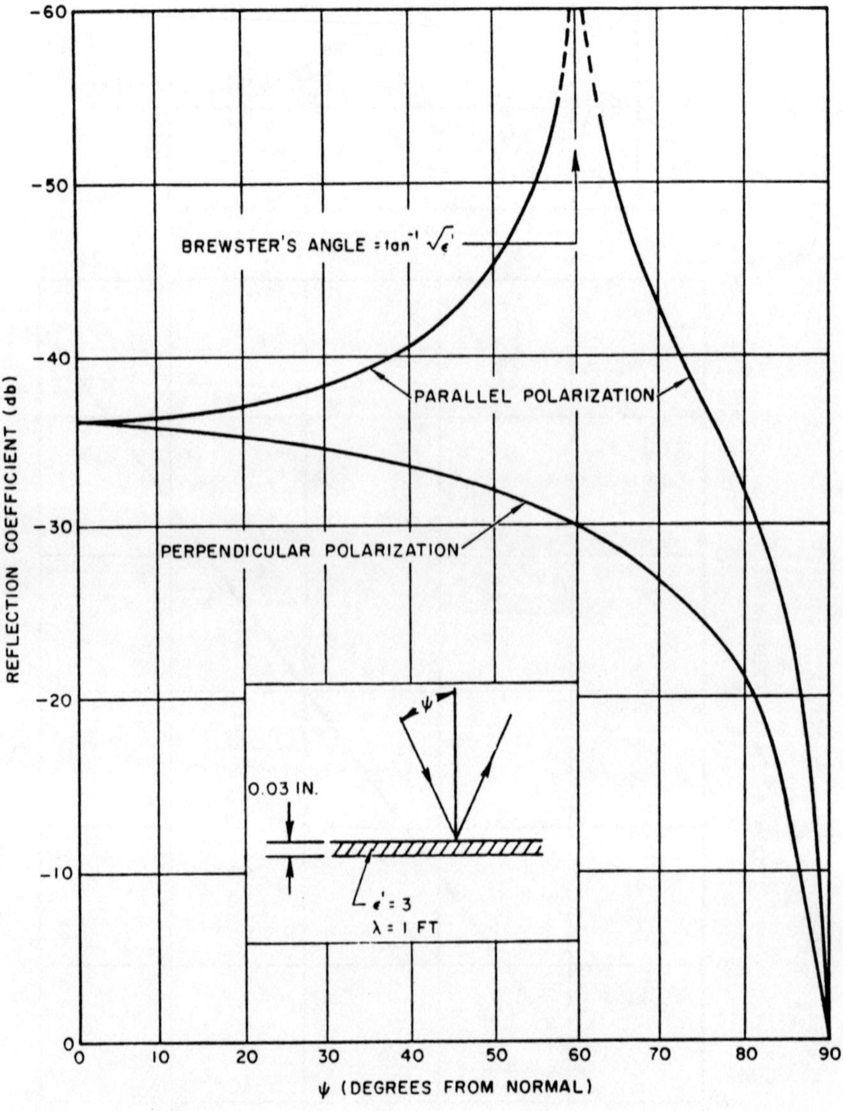

Figure 4–9. Reflection Characteristics of Radome Material for Perpendicular and Parallel Polarization

background level. As an alternative to the exceedingly bulky and expensive anechoic chamber, it might be possible to house a range in a large air-supported plastic radome several hundred feet in diameter. The parallel and perpendicular polarization reflection characteristics for a typical radome material are shown in figure 4–9. The reflection from walls is −36 dB at normal incidence. Since a curved surface admits many reflections at angles other than normal incidence, it seems improbable that a radome can be designed to reduce all reflections to −35 dB. Because this level may be unsatisfactory for small-cross-section targets, the radome range seems unsuitable for accurate measurements.

Conventional Outdoor Range

Most of the full-scale ranges, both for measuring antenna patterns and for measuring scattering cross-section, are of the outdoor type. Here a transmitting/receiving antenna and a target-support tower are located the appropriate distance apart on some convenient patch of ground. This type of range has three inherent drawbacks. First, in contrast to the anechoic chamber, the entire physical situation is not under the control of the experimenter. Effects of rain, humidity, temperature, vehicles, and wind may be appreciable and cannot be readily—if at all—controlled. Second, the outdoor range admits a significant specular contribution to background: in addition to the direct ray between antenna and scatterer, there is a ground-reflected ray that is essentially specular. The serious field perturbation produced by this ground reflection has been the bane of pattern range designers for many years. Third, the outdoor range admits a variety of nonspecular reflection paths. For example, unless the terrain between antenna and scatterer is exceedingly flat, smooth, and level, energy will be reflected between antenna and scatterer from many points in the region between the two. In addition, reflection and diffraction occur from any earth mounds, towers, hillsides, and so on used to increase the height of antenna and scatterer above the terrain. Scattering of the incident wave distorts the field in which the target is immersed. In addition, bistatic scattered energy from the target may be diffracted by the target support into the receiving antenna.

When the environmental factors are controllable, as in an anechoic chamber or a radome-covered range, those factors that are not a function of the target can be reduced by a background-cancellation technique. Although background cancellation can be used in outdoor ranges despite changing parameters, the general background obtainable is poor, and repeatability of data is also often poor. To give an idea of the magnitude of the specular ground-reflected wave, figure 4–10 shows the reflection coefficient for average earth at L-band. These curves are intended to give only order-of-magnitude effects, as the great variety of situations precludes any direct use of the data. However, from figure 4–10 it may be observed that except for the

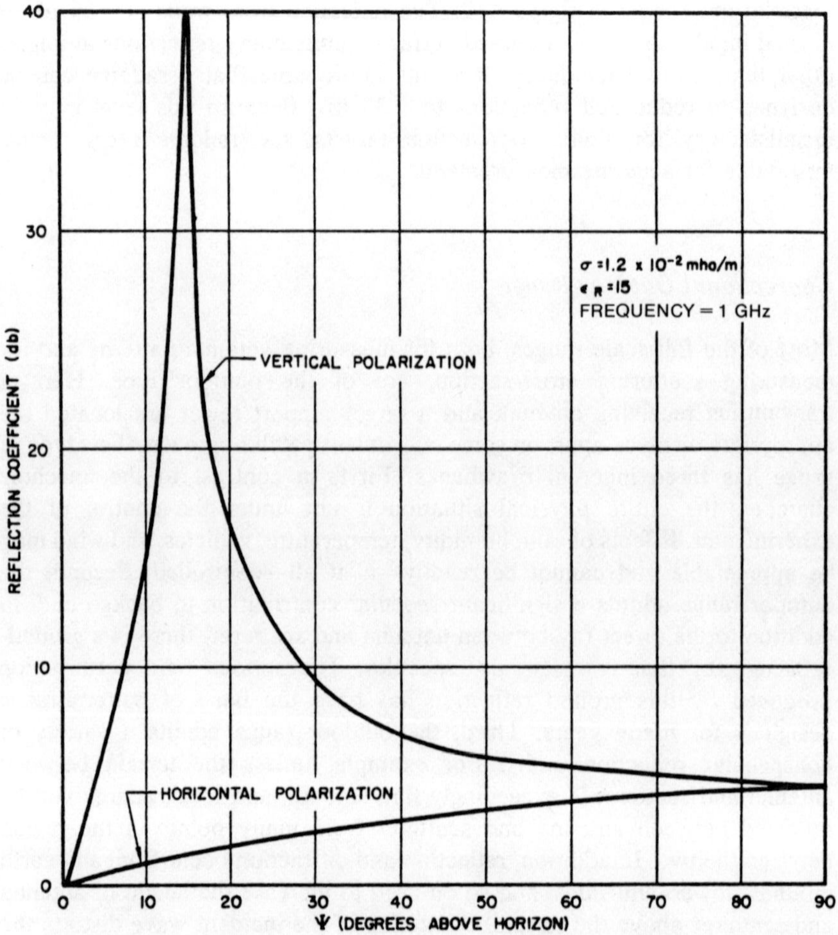

Figure 4–10. Reflection Coefficient for an Average Earth at 1 GHz

region near the Brewster angle of incidence, the reflected wave is likely to be comparable in magnitude to the direct wave.

To minimize the reflection characteristics of the ground, the field is often plowed. Except for relatively high angles of incidence, plowed fields provide essentially no attenuation to the ground-reflected wave. Some insight into typical earth-reflection coefficients can be obtained from figure 4–11, which gives data for plowed fields at S-band. Data are given for propagation normal to the furrows and at 45°, and for both horizontal and vertical polarization [18]. Here again it may be noted that, even allowing several bounces, the attenuation is nowhere near that needed for low-cross-section results.

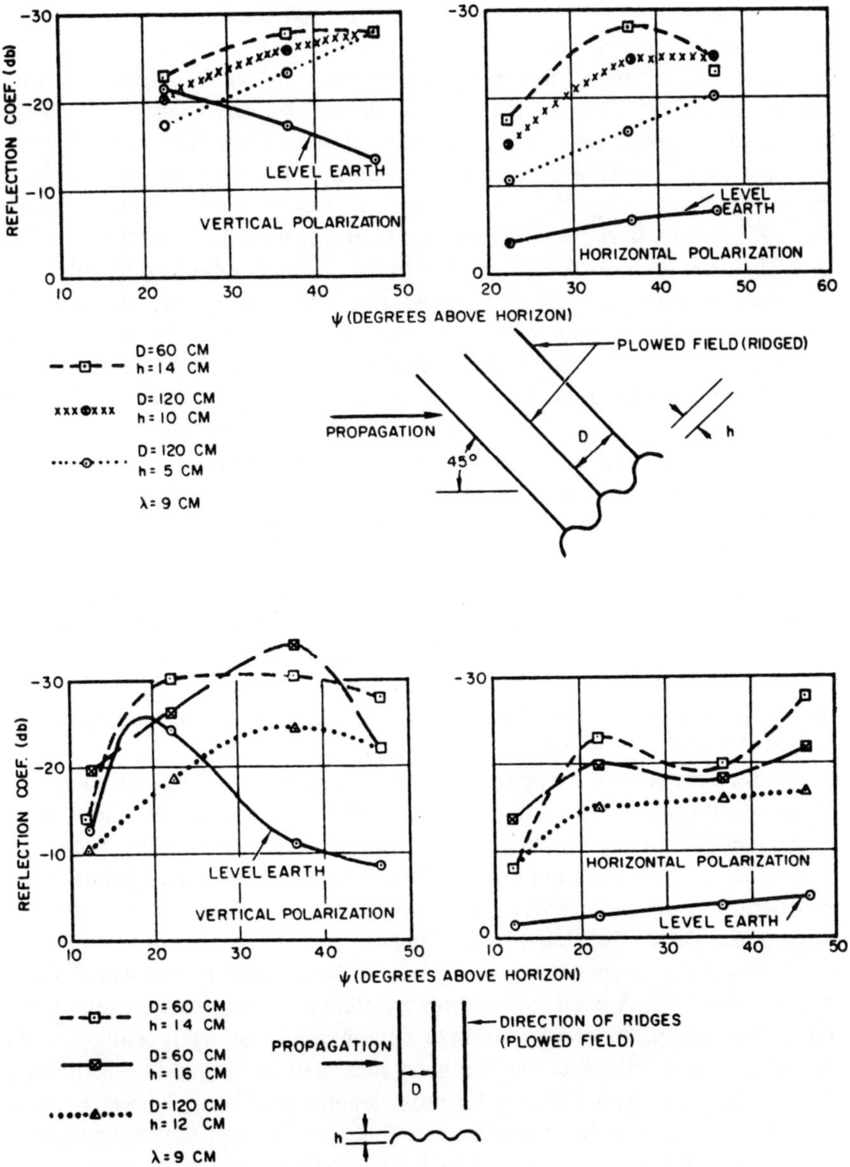

Figure 4–11. Reflection Characteristics of Plowed Fields

The ground-reflected ray can be intercepted by the use of diffraction screens, which are often referred to as *radar fences*. The geometry is shown in figure 4–12, where for this case the screen is located midway between source and sink. This location is not unique by any means but, rather, represents a typical performance regardless of location. The interference of the direct wave and the diffracted wave produces the Cornu spiral found in treatises on knife-edge diffraction [19]. The resultant field strength derived from the Cornu spiral is plotted in terms of a generalized variable v, where $v = 2h/\sqrt{d\lambda}$, in figure 4–12, which shows that the field falls off rapidly when the source and the sink are below the knife edge and produces the interference pattern when the field of interest is above the knife edge. In order to minimize the field-amplitude variations, the height above the knife edge must be sufficiently great. For ease in evaluting v as a function of height and range, the nomograph of figure 4–13 is useful. For example, if a maximum amplitude variation of ± 0.5 dB is desired, a v in the range of 3.8 is indicated. From the nomograph, this would require a height of 23 ft at a distance of 300 ft, or 16 ft at the 150-ft range. This determines the height of the target axis above the fence. To determine the variation over the target height (or width), the nomograph in figure 4–14 can be used to determine the excursion of v about the on-axis value. For example, at a point 1.5 ft above or below the 23-ft height mentioned earlier (for a 300-ft range), a change in v of 0.13 is produced. By reference to figure 4–12, it is found that the 0.13 excursion allows the maximum amplitude variation along the height of the target to be almost 0.5 dB.

Radar fences would appear to solve the problem of the specular ray. However, it often happens that a number of fences are needed and that the placement of these is exceedingly crucial with regard to frequency, atmospheric conditions, antenna beamwidth and pointing, and polarization. Furthermore, the radar fences may produce new background contributions through the waves diffracted by the fences.

Some of the extraneous background can be reduced by the use of short-pulse radars [17]. A good rule of thumb is that a 2-nsec pulse will give 1 ft of target resolution. In order to utilize narrow-pulse gating to eliminate the specular ground reflection, the geometry must satisfy the conditions of figure 4–15, where curves are shown for pulse lengths of 25 and 5 nsec. Here an ideal range gate has been assumed, and it can be seen that for ranges of several hundred feet in length, the height required is impractically large. However, short-pulse radars are certainly useful for reducing the return from buildings and other obstacles with range appreciably different from that of the target. The minimum length of the pulse is limited by several factors, including those involving the electronic equipment. For example, a 25-nsec pulse duration requires a bandwidth of the order of 50 MHz; with a carrier frequency in L-band, conventional tubes such as magnetrons may not be

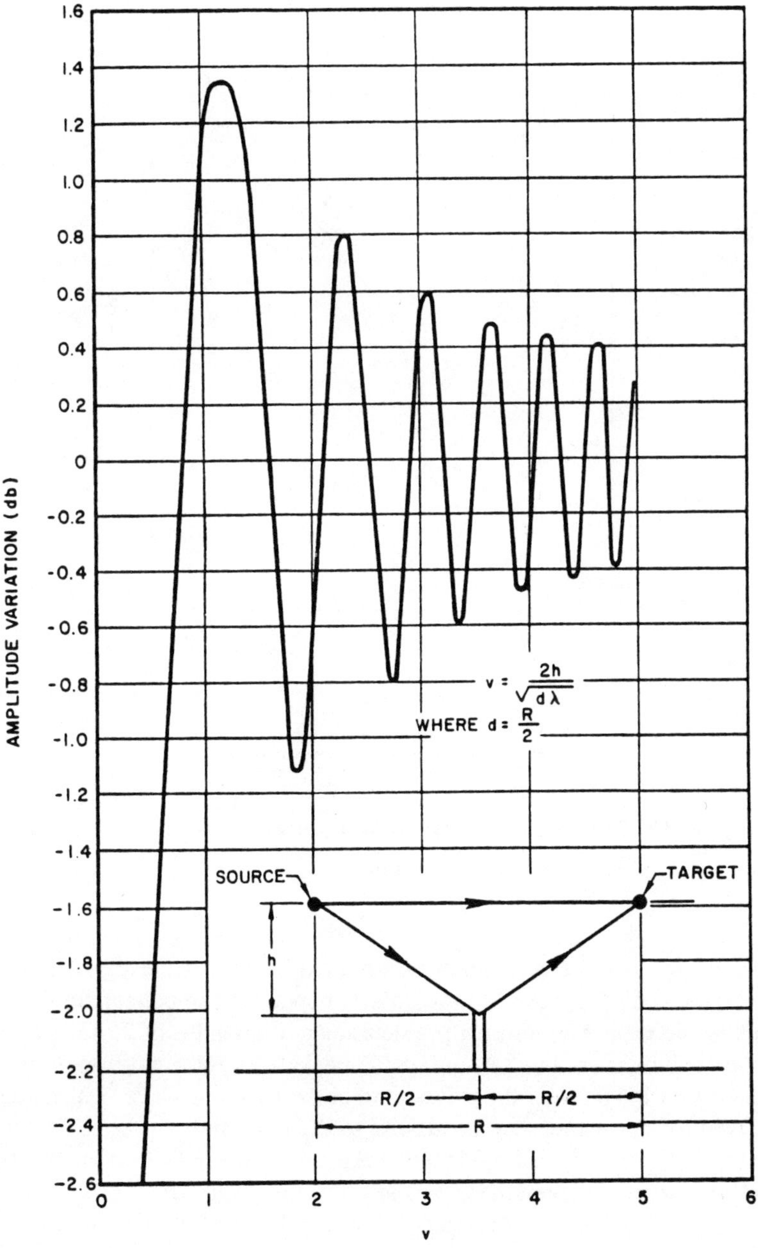

Figure 4–12. Amplitude Variations Caused by Knife-Edge Diffraction

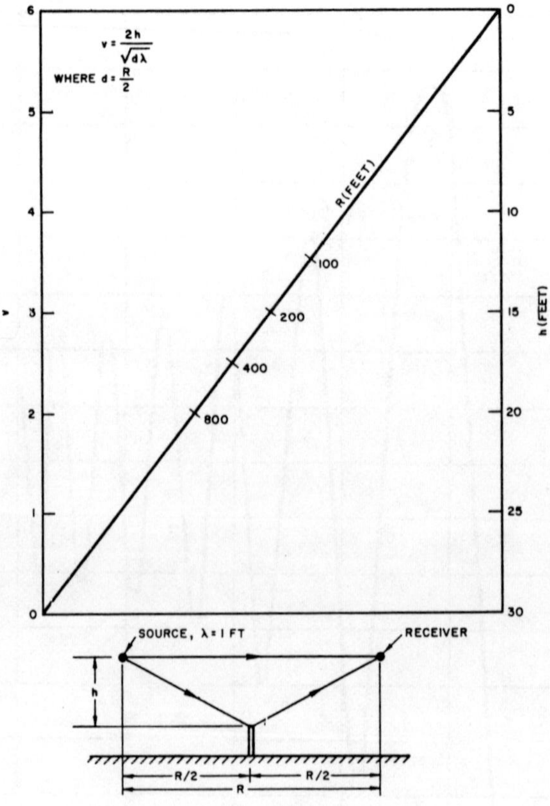

Figure 4–13. Nomograph for Evaluating v as a Function of h and R for Knife-Edge Diffraction

satisfactory. Another limitation is provided by the frequency sensitivity of the target and background to the different spectral components in the pulse. This would probably not be a problem for pulse lengths of 25 nsec or more. Another limitation is that the pulse-packet length (in time) should be appreciably longer than the travel time of the traveling wave, which starts at the front of the scatterer, goes around to the rear, and returns to the front of the body. For a body of 10-ft length, taking into account a modest obtuseness and the lower velocity of the traveling wave, one might arrive at an equivalent free-space length of 25 feet; thus pulse lengths of at least several times 25 nsec are indicated. Unfortunately, these pulse lengths limit the background suppression obtainable by this technique, making a design compromise necessary.

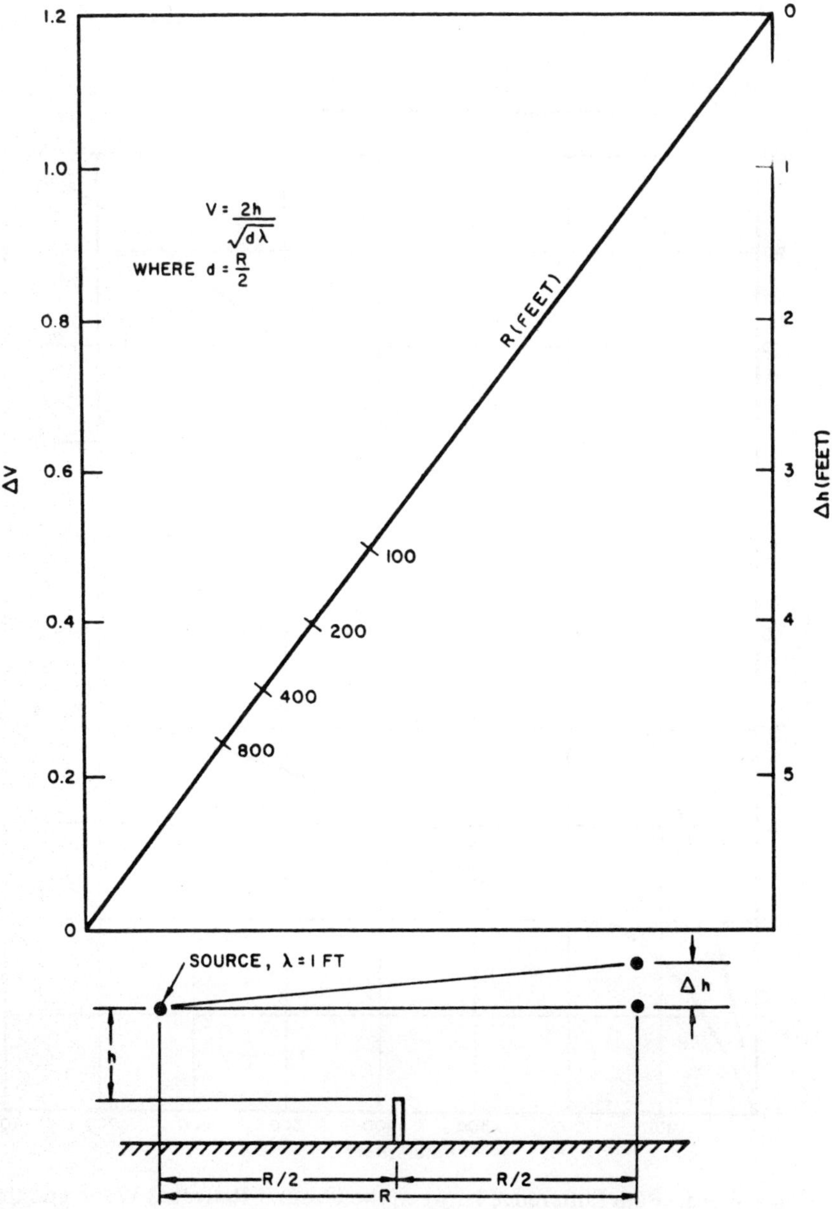

Figure 4–14. Nomograph for Evaluating Excursion of v as a Function of Target Height and Range for a Knife-Edge Diffraction

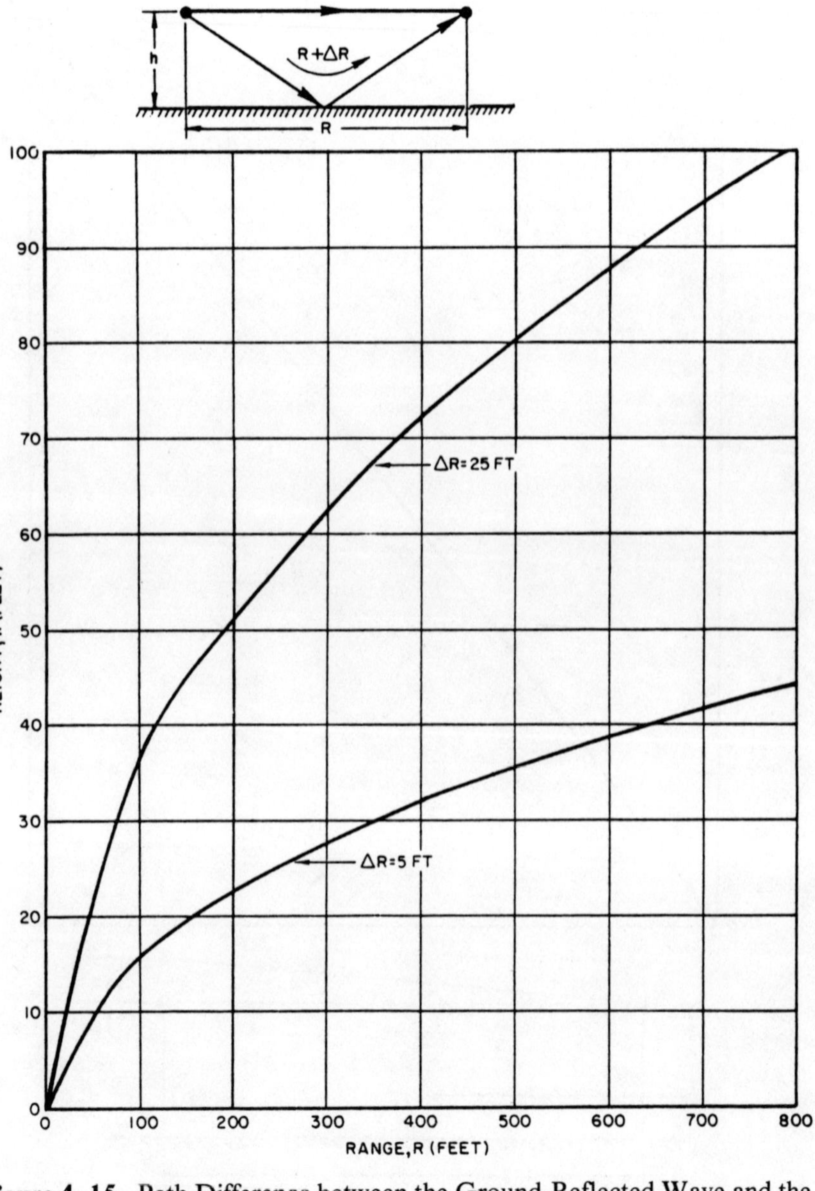

Figure 4–15. Path Difference between the Ground-Reflected Wave and the Direct Wave

The Ground-Plane Range

From the previous discussion on ground reflections, one may conclude that the random reflections from the intervening ground preclude accurate measurements. All available data indicate that these random reflections must be highly suppressed. The ground-plane range and the oblique range, to be discussed next, both minimize the random reflections. In the ground-plane range, the ground is made flat and smooth with respect to λ. The antenna and scatterer are located in that part of space where the direct ray and the ground-reflected ray produce a maximum. This type of range has been used by Lincoln Laboratories [20], Armour Research Foundation [21], Bell Telephone Laboratories [22], and others for measuring antenna systems. For example, Lincoln Laboratories have measured sidelobe nulls better than 60 dB below the main beam or about 17 dB below an isotropic radiator level; and this by no means represents the limit in performance. The direct and reflected rays interfere to produce a cosine pattern, the maximum of which occurs at the height given in figure 4–16. From this figure it can be seen that a 200-ft range requires the antenna scatterer axis to be 7 ft above the ground. Figure 4–17 depicts the amplitude variation as a function of target height in feet; and, for a displacement of 1.5 ft from the maximum point, the 200-ft range produces only an 0.5-dB amplitude decrease.

The Lincoln Laboratories antenna range is 2,000 ft in length and has a ground surface maintained within one inch of flat and smooth, even during the severe winters common to the Boston area. Contrary to the Lincoln Laboratories technique of using a smooth ground, the Armour Research Foundation used an unconditioned ground.

A study of the feasibility of using the Lincoln Laboratories ground-plane antenna-range technique for cross-section measurements has yielded high promise [36]. By eliminating grass blades, which contribute significantly to the background cross-section, and by limiting the lower operating frequency to comply with the requirement of placing the peak of the interference pattern at the center of the target while obtaining nearly equal ground reflection coefficients for vertical and horizontal polarizations, it may be possible to obtain residual background levels (due to the ground backscattering alone) on the order of -75 dBλ^2 for pulse durations of a few tens of nanoseconds. The example given is for a concrete surface.

The need for circular polarization data necessitates special considerations of the ground-plane range. A circularly polarized wave can be resolved into two orthogonal components (say vertical and horizontal) with a 90° phase difference. The difference in the magnitude and phase of the vertical and horizontal ground-reflection coefficients as a function of grazing angle can therefore distort the composite field illuminating the target. Since the

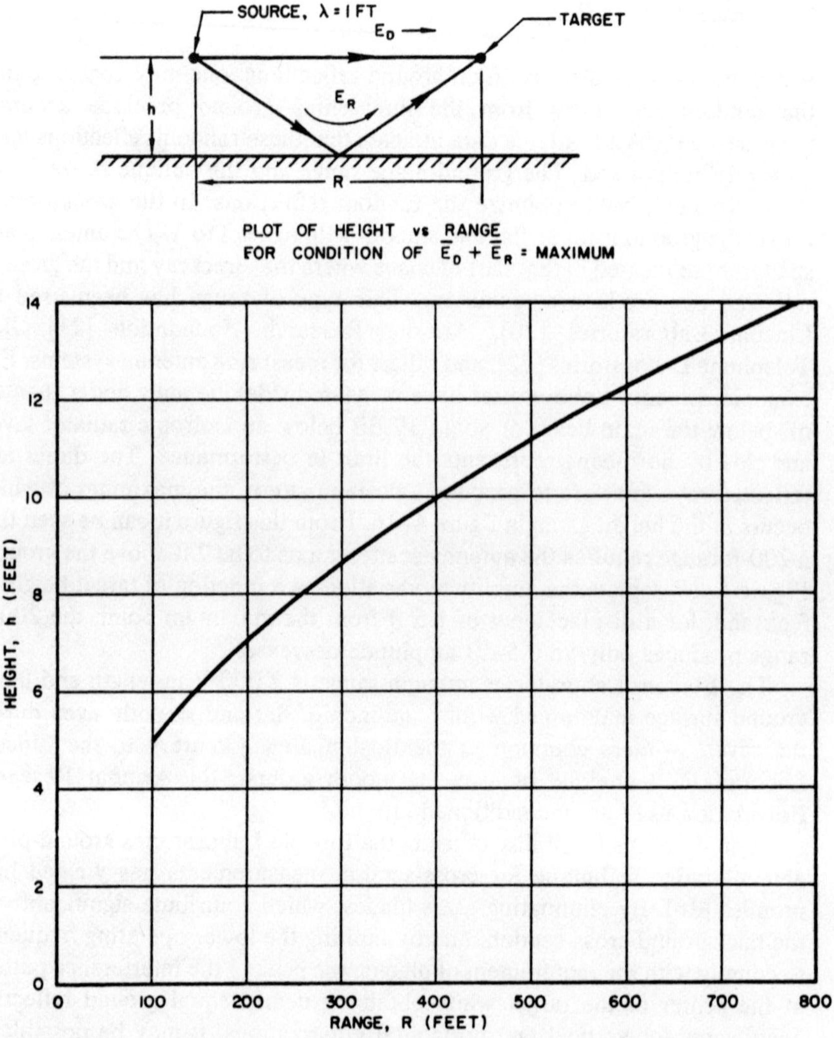

Figure 4–16. Height and Distance Required for a Ground-Plane Range

ground-plane geometry is fixed by consideration of frequency, antenna aperture, and target size, it is not always possible to minimize grazing angles. One solution might be to use a turnstile junction, modified with suitable attenuators and phase shifters, as a primary feed for the illuminating antenna to transmit a "predistorted" elliptically polarized wave such that the composite wave at the target is correct [38, 39]. A calibration target with a

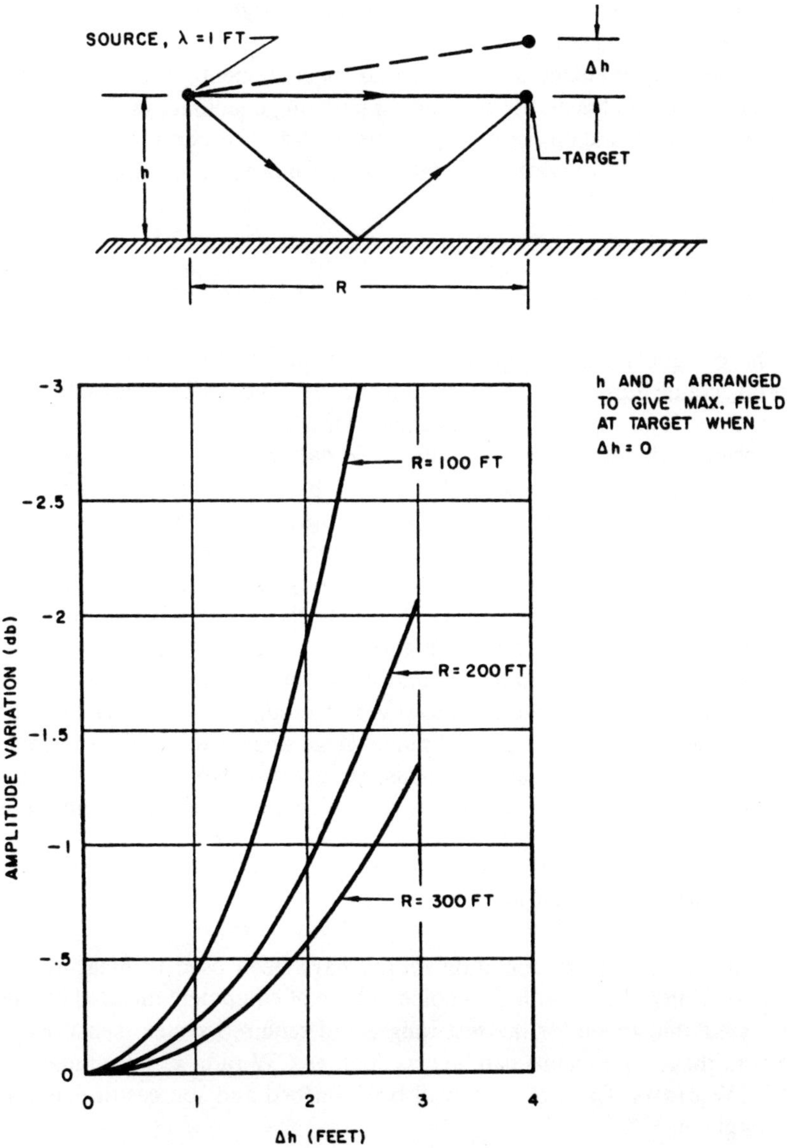

Figure 4–17. Amplitude Variation over Target in a Ground-Plane Range

known response to circular polarization could be used to calibrate the entire range in this fashion [38, 40].

If the grass is removed (water-filled grass blades correspond to scattering dipoles at frequencies of interest) from a ground-plane range, it is estimated that the residual background cross-section (no target supports) will be -50 $dB\lambda^2$ to $-60dB\lambda^2$ [36].

A carefully conducted experimental measurement program will determine the ultimate feasibility of ground-plane range techniques. In addition to measurements made on earth, airport runways and/or parking lots constitute likely candidates for gathering data on ground-plane cross-section range techniques.

Oblique Range

In the oblique range the ground return is minimized by simulating more closely actual target-detection circumstances. The radar is located on the ground and looks up at an oblique angle at the target; the antenna beamwidth is chosen so that the ground is illuminated only by low-level sidelobes. This technique has been used very satisfactorily by Dawson and Edwards at the Royal Aircraft Establishment [16]. They suspended the target by nylon rope. One disadvantage of this system is that as the rope rotates, a skew pattern is obtained from the target. However, for targets consisting of bodies of rotation, the skew pattern can be converted to the ordinary longitudinal-plane pattern by a nonlinear transformation of variables. The range heights necessary to achieve suitable ground-illumination control make this technique somewhat less attractive. However, it should be emphasized that both the oblique range and the ground-plane range appear to offer a satisfactory solution to the low-cross-section measurement problem, whereas the ranges discussed in radar cross-section range problems have intrinsic limitations.

Equipment Considerations

Several types of instrumentation radars have been used to measure radar cross-section [14,23], with the choice of type of equipment dictated by target size, available space for the test range, and required system sensitivity. In general, these equipments can be classified as CW radars, pulse radars, and FM/CW radars. Each of these will be described and sources of equipment errors given.

CW and MTI Radars

A CW radar of the bridge type is shown in figure 4–18 [1,23,24]. RF power is injected into the H arm of a magic T, with the scattered signal appearing in

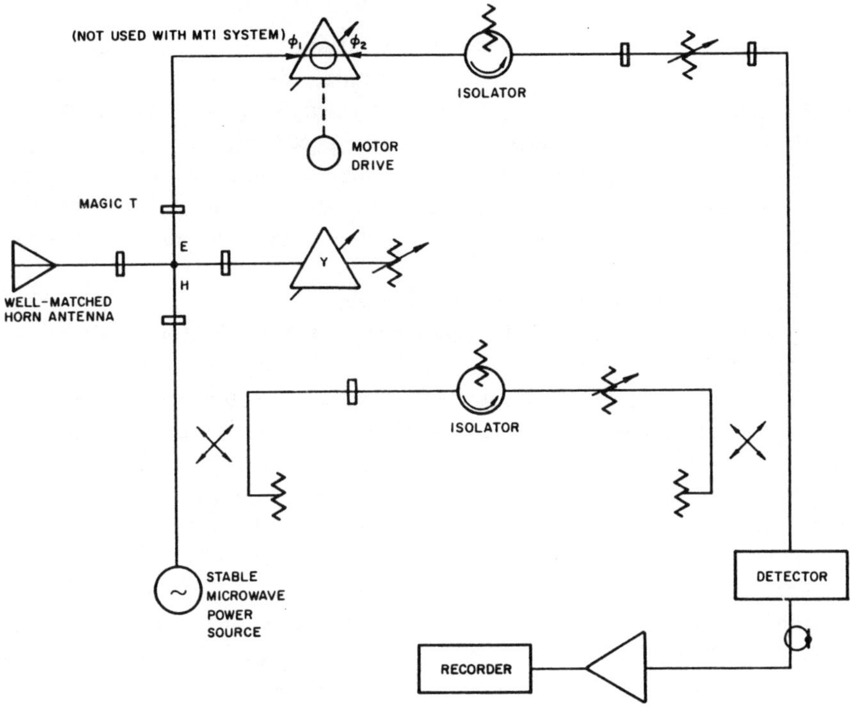

Figure 4–18. Illustrative Block Diagram of CW Bridge Radar

the E arm of the magic T.[8] If this signal in the absence of a target is small, it can be nulled out by adjusting the side-arm termination without affecting the linearity. The actual target then produces a signal in the E arm; this signal is amplified in the receiver. The target can affect the background signal and thus its cancellation; this consideration may be imporant for large targets that have low cross-section.

Direct conversion of RF to audio can be effected by use of a motor-driven phase shifter, as shown in figure 4–18. Synchronous detection may also be utilized by detecting the amplified signal in a phase detector with a reference signal from the motor-driven phase shifter.

The measurement procedure is straightforward and can be made nearly automatic (with additional equipment) once calibration has been effected. A known target such as a sphere gives a reference indication and is replaced by a target of unknown cross-section. The precision attenuator reading is noted for each target with the same indicator reading, giving radar target cross-section in dB referred to the object of known cross-section. To automate the process so that a continuous record of cross-section versus aspect angle can be made, a motor-driven attenuator can be used as part of a control loop to null the indicator output as the target is rotated. the servo error signal and aspect angle signal from the target are recorded on a strip chart, and data reduction yields radar cross-section versus aspect angle [25].

When the frequency is high, the distance needed is small, making the employment of a compact indoor CW range attractive (since D/λ is usually fixed, $R \sim \lambda$) [12,14]. Such a range is more suitable for measuring scale models than for full-size targets.

An improvement in background suppression can be made by using moving target indication (MTI) techniques [11]. Here the target is rotated, nutated in a circle, translated, and so forth. In case of periodic relative motion, the scattered signal can be coherently detected with a reference signal, affording additional sensitivity. The nutation method has been used with success by Scharfman and King [26] and others [27,28,29]. A limit occurs for large scatterers because the attainable rotation or nutation speed is much lower than that for small bodies; a speed of 3,000 rpm corresponds to an audio frequency of only 100 cps because of changes in return from the antenna pattern, so that even at this speed, selective filtering is difficult. Another scheme moves the target slowly in the axial direction [30]. Cross-section is then plotted versus range, and an average $1/R^4$ curve is fitted. Another technique, suitable for measurement of backscatter from antennas, uses a photoconductive termination on the antenna, which is then periodically modulated by a light source. Still another bridge-type scheme uses a Michelson interferometer as the bridge [28].

The ultimate sensitivity of the CW radar is limited by imbalance of the various signals, rather than by the thermal noise. The narrow bandwidths possible with frequency-stabilized transmitter and phase-locked local oscillator bring about this situation. Thermal expansion of the magic T junction, target-location effects on background, and difficulties in matching and alignment of the tunable termination all tend to make this type of radar difficult to use with other than a restricted class of target-wavelength combinations.

Pulse Radars

A pulse radar does not have the coherent background-cancellation attribute of the CW systems (unless the radar is coherent) but can discriminate against some background objects by use of a sharp range gate [31]. Since a 1-nsec pulse corresponds to a range resolution of 0.5 ft, use of very short pulses can afford excellent discrimination. Unfortunately, as mentioned previously, the target must be immersed in essentially a CW field, so that the travel time of the diffracted energy around the scatterer is short with respect to the pulse time. This severely limits the range-rate discrimination that can be used. Such a radar is not sufficiently different from conventional radars to warrant a block diagram. Several points may be mentioned, however. If the distance to target is sufficiently large, separate receiving and transmitting antennas may be used to eliminate duplexing problems. Note that separate antennas used in this fashion constitute an approximation to the monostatic case, becoming better as the ratio of target range to antenna separation becomes larger. One method of eliminating the recovery-time problem of conventional duplexing devices is to use a solid-state (varactor) duplexer to allow use of a single antenna for transmitting and receiving. The solid-state duplexers can duplex a few tens of kilowatts in peak power but have bandwidth limitations. It will be shown that for most distances very low powers are sufficient, so that the usual magnetron or klystron output may not be needed. The CW cancellation advantages could be realized by use of a coherent radar. This requires generation of a series of phase-coherent harmonics, which are properly assembled to constitute the pulse. The returned echo is then phase-coherently detected, and a background cancellation can be effected here. This system is similar in some respects to the coherent-pulse-doppler radars [32], but it remains to be seen how much advantage can be gained for backscatter measurements.

Short-pulse radars with pulse lengths as short as 1 nsec have been used to obtain a partial return from a specific portion of the target—for example, from the tip of a long cone. Because there is considerable uncertainty about the effects of transients on such scatterers, no further discussion is warranted here.

FM/CW Radar

The basic principle of the World War II FM/CW radar altimeter [32,33] has been used by CAL and others [34,52] for the isolation of signals arriving from a desired range interval. In this scheme, a repetitive linearly increasing (sawtooth) frequency variation is transmitted; by comparing the received frequency with a reference, it is possible to filter out selectively all reflected energy not coming from a range "block" around the target. In practice, to minimize effects of base-frequency shift, the reference signal is sent through a delay line (waveguide or free space) of electrical length nearly equal to the path length from and to the target. The received signal and reference are then heterodyned and filtered. Figure 4–19 is an illustrative block diagram. Amplitude weighting of each period of the signal is used to reduce "sidelobes" in range-space.

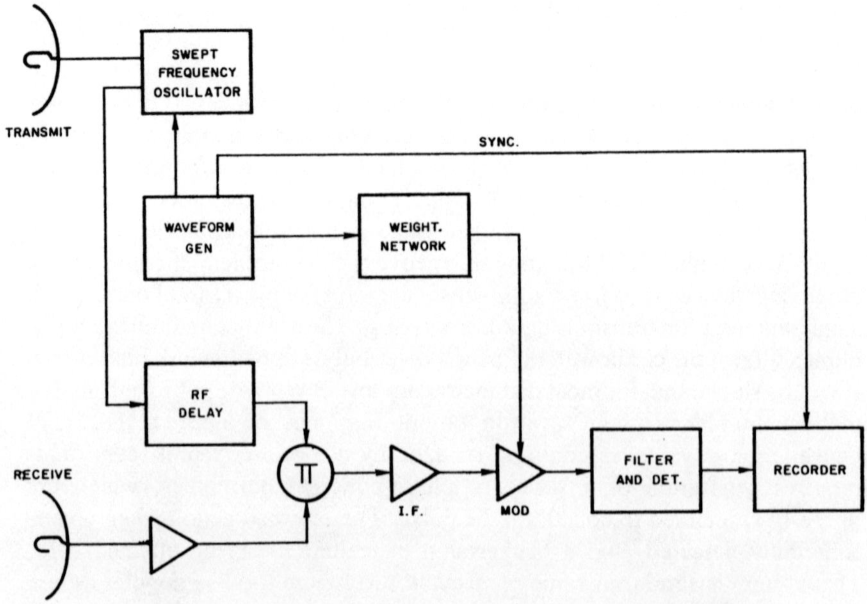

Figure 4–19. Illustrative Block Diagram of FM/CW Radar

A limitation of this approach is inherent in the phase change with time of the illuminating wave. Even a small frequency change during the transit time of the slow wave around the target will produce a very large phase shift. This technique therefore appears limited to physically small targets.

Errors and Calibration

There is an appreciable literature on the probability of detection of signals in noise, for various waveforms and types of signal processors. The translitera-tion of these results to the problem of determination of cross-section from automatic gain control (AGC) or other measured data is not within the scope of this chapter. It will suffice to observe that for stochastic processes it is neccessary to specify both an error magnitude and the probability of occurrence of that or a lower value. For simple systems the normal distribution applies, and there is a direct correspondence between standard deviation (σ) values, and percentage confidence factors.[9] The correspond-ence is well known but is repeated here for convenience.

Deviation	*Confidence Factor(%)*
$\pm 1\sigma$	68.3
$\pm 2\sigma$	95.5
$\pm 3\sigma$	99.7

Thus, if it is desired to measure cross-section to ± 1 dB within $\pm 2\sigma$ limits, 95.5 percent of the time the cross-section error will be less than ± 1 dB.

One of the most crucial phases of operation is the calibration phase. Typically, the AGC level of the radar is recorded, with one calibration point being obtained using a known target. This process is not good enough for very small targets. The following calibration procedure is necessary but not sufficient for measurement of low cross-sections from large shapes. First, the AGC range of the radar receiver is adjusted so as to cover the desired cross-section variation.[10] Then the AGC is calibrated by use of a sphere whose cross-section is near the upper end of the variation. To do this it is convenient to have curves of cross-section with respect to λ^2 versus sphere diameter, d/λ. Such curves have been plotted from exact scattering data and are shown in a normalized form in figure 4–20.[11] Once the AGC has been calibrated and a point plotted on a sheet of graph paper, the cross-sections of a series of spheres down to and including a sphere representing the low end of the AGC range are measured. Results from each sphere are plotted on the graph sheet, along with probable error as a vertical line. The calibration curve then shows the linearity of AGC and whether the facility can in fact resolve targets at the low end of the variation. The upper portion of the sphere data curve exhibits resonance oscillations about the geometric optics value of $\pi d^2/4$, and the lower region is asymptotic to the Rayleigh value of $9\pi^5 d^6/4\lambda^4$ [49].

The conversion curves of figure 4–20a are available for convenience in converting the normalized sphere data to the cross-section in square meters at a specified frequency. (Add the dB values from figure 4–20a to the dB values of figure 4–20b to obtain the sphere cross-section in dB m^2.)

The foregoing calibration procedure is an exceedingly difficult test for an outdoor radar cross-section range; many ranges with a 60-dB AGC variation are not able to discriminate targets at the lower end. However, these sphere tests are not sufficient, as it is possible for a carefully balanced cross-section range to measure spheres accurately but to have the balance disturbed by the presence of a large scattering object such as a cone shape. In order to check the accuracy in this respect, it will be necessary to develop a set of physically large shapes of known low cross-section for calibration purposes—for example, dielectric spheres, thin-metal film-coated dielectric spheres, and possibly a set of calibrated cones.

Preliminary analytical and experimental data on dielectric spheres have

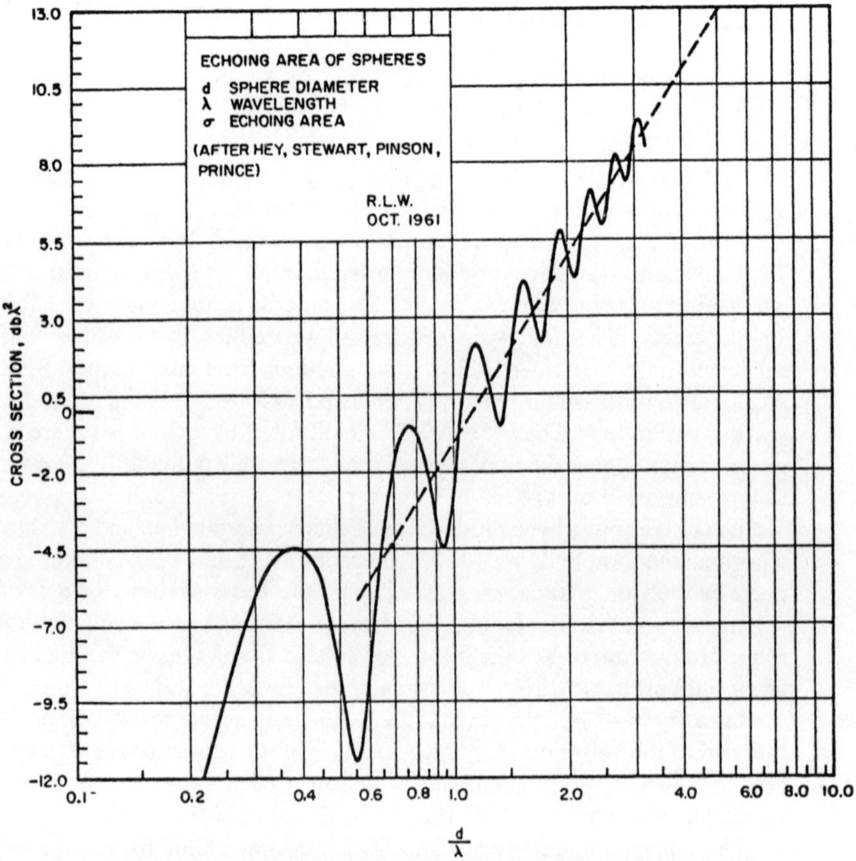

Figure 4–20. Ratio of Backscattering Cross-Section with Respect to λ^2 for a Conducting Sphere with a Diameter d/λ

recently become available. Garbacz has presented bistatic curves of cross-section as a function of sphere size and the sphere's intrinsic impedance [43]. Adler, with the use of a pulsed tracking radar, made backscattering measurements on 120 dielectric and metal spheres at 5,518, 5,600 and 5,800 MHz [44]. Adler reveals good agreement with Mie's theory in the critical size region for Teflon, Lucite, Bakelite, and aluminum spheres.

Because low-cross-section radars are very sensitive to environmental changes such as rain, insects, animals, and foliage, it is probably not possible to provide continuous capability of operation. Shutdowns due to snow, rain, floods, large diurnal temperature changes, insects and birds, and so on will be necessary unless some type of climatic control is provided.

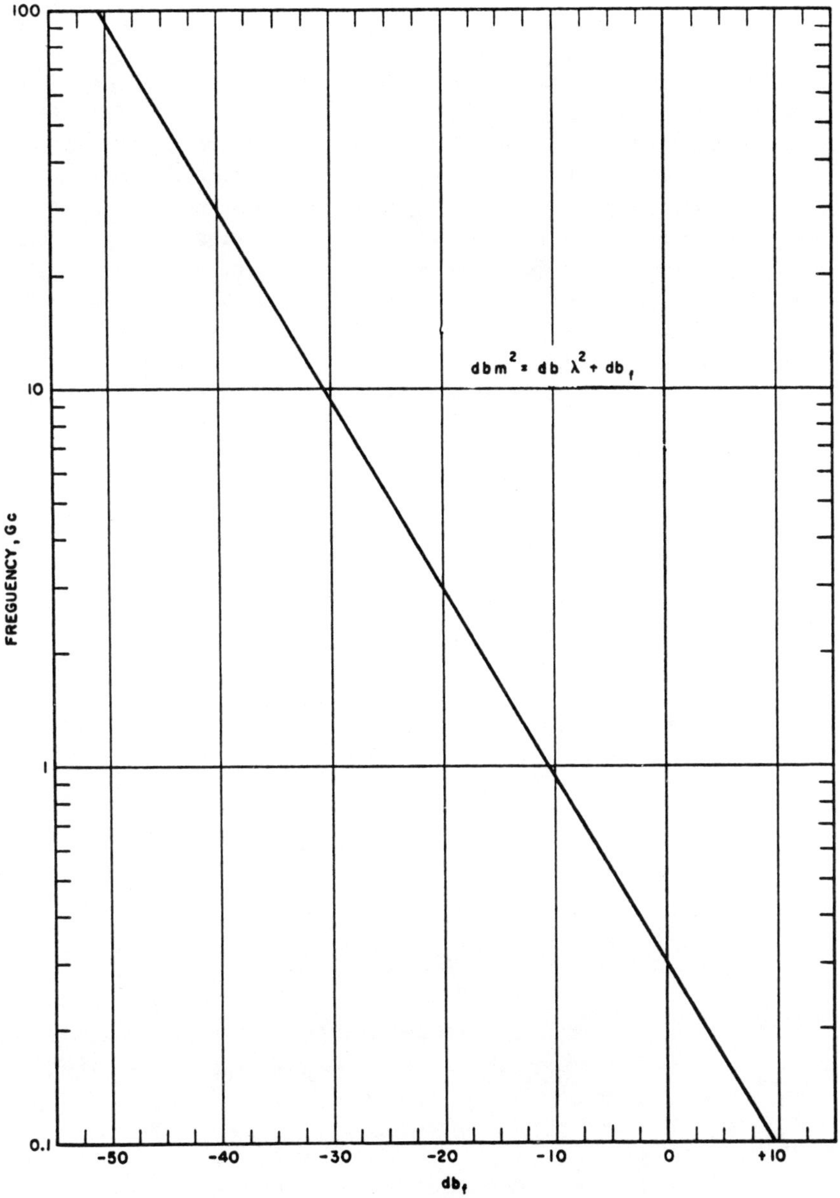

Figure 4–21. Curve for Converting Cross-Section in dB λ^2 to dB m^2

Conclusions

The achievement for a long-distance cross-section range of sensitivities of the order of $10^{-2}\lambda^2$ to $10^{-4}\lambda^2$, which are now realized with short indoor anechoic-chamber cross-section ranges, is a formidable engineering task. The ground-plane and oblique ranges appear to offer good possibilities of realizing or bettering this goal. A short-pulse radar with pulse lengths of the order of 100 nsec is desirable and could be made coherent in order to allow phase cancellation of some background. Target supports might consist of polyfoam and/or absorber-covered columns for medium-cross-section targets, and plastic rope for low-cross-section targets.

Using the ground-plane range geometry with a radome-balloon target protective structure, both short-pulse and continuous-wave RCS measurements were obtained.

Range Equation

The passive radar range equation can be written as

$$P_{\text{trans}} = \frac{64\pi^3 R^4 P_{\text{rec}}}{G^2\lambda^2\sigma} = \frac{64\pi^3 R^4 \dfrac{S}{N} kTB}{G^2\lambda^2\sigma}, \tag{4.14}$$

where k = Boltzmann's constant ($k = -198.6$ dBm, per cycle per degree),

T = system noise temperature in degrees K,

G = antenna directivity (often called gain),

B = bandwidth.

Typical paraboloid aperture gain factors are 50 percent, so

$$G = \frac{\pi^2 D_{\text{ant}}^2}{2\lambda^2}. \tag{4.15}$$

Assume $D_{\text{ant}} = D_{\text{tar}}$, and also assume that the distance R is chosen as the sum of $2D_{\text{tar}}^2/\lambda$ and $1.5\ D_{\text{ant}}^2/\lambda$. So

$$R = \frac{3.5\ D^2}{\lambda}. \tag{4.16}$$

Substitution of these gives

$$P = \frac{12{,}200 \dfrac{S}{N} kTB\, D^4}{\lambda^4 \sigma/\lambda^2}, \tag{4.17}$$

where σ/λ^2 has been separated as the quantity of major interest. Evaluation of the constant produces two results:

$$P\tau \frac{\sigma}{\lambda^2} = \frac{S}{N} TD^4 f^4 (-127.5 \text{ dB}), \tag{4.18}$$

$$P\tau \frac{\sigma}{\lambda^2} = \frac{S}{N} T \frac{D^4}{\lambda^4} (-127.7 \text{ dB}). \tag{4.19}$$

The number in parenthesis is the resulting constant multiplier expressed in dB; τB has been assumed unity.

In both equations, the units are

P = transmitter power in watts,

S/N = signal-to-noise ratio,

τ = pulse length in μsec,

T = system noise temperature in degrees K.

In equation 4.18 the frequency f is in GHz, and the antenna diameter D is in feet. D/λ in 4.19 is dimensionless, as is σ/λ^2 in both equations.

Some typical assumptions may indicate the power levels involved. Assume a single pulse S/N of 20 (13 dB); crystal video receiver with overall temperature of 2,910°K (noise figure = 10 dB); target of $D = 20\lambda$. Then,

$$P\tau \frac{\sigma}{\lambda^2} = -28.5 \text{ dB}. \tag{4.20}$$

A plot of this is given in figure 4–22. It is apparent that ample margins can very easily be provided because of the relatively short ranges.

To these data must be added more careful consideration of system losses, of values of $\tau B > 1$, and of what S/N are needed to yield results within either a prescribed percentage confidence level or within multiples of a standard deviation.

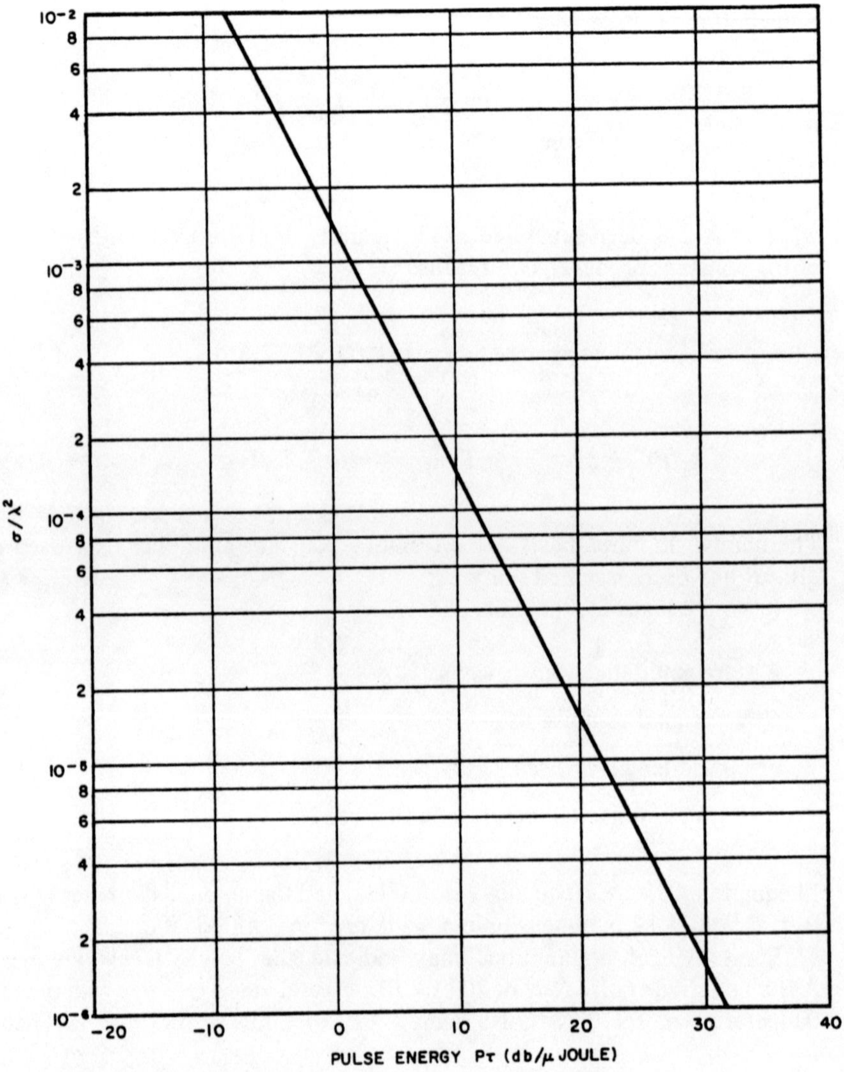

Figure 4–22. Pulse Energy for $S/N = 20$, $D = 20\lambda$

Pulse Integration

Figure 4–22 indicates that high powers are required at the higher frequencies. Pulse integration is one method to reduce the required transmitter power.

The minimum interpulse period for a known range from the radar to the target is T'.

$$T' = 2R/c. \tag{4.21}$$

The corresponding *prf* is

$$\frac{1}{T'} = \frac{c}{2R}. \tag{4.22}$$

The range distance R may be expressed as

$$R = p\frac{D_{\text{tar}}^2}{\lambda}, \tag{4.23}$$

where p, usually in the region of 0.5–4, is chosen on the basis of the required fine structure of the cross-section pattern.

The *prf* in terms of p is

$$1/T' = c\lambda/2pD_{\text{tar}}^2. \tag{4.24}$$

Equation 4.24 gives pulse repetition rate as a function of other radar cross-section range parameters, including frequency and maximum projected target dimension.

If the cross-section does not change appreciably in some incremental aspect angle $\Delta\theta$ and the angular velocity $\dot{\theta}$ is constant, the incremental time is

$$\Delta t = \frac{\Delta\theta}{\dot{\theta}},$$

and the number of pulses integrated is

$$N = \frac{\Delta t}{T'} = \frac{\Delta\theta}{\dot{\theta}} \, \frac{c\lambda}{2pD_{\text{tar}}^2}. \tag{4.25}$$

With the target size, D_{tar}, expressed as a multiple of a wavelength

$$D_{\text{tar}} = k\lambda,$$

then the number of integrated pulses is

$$N = \frac{\Delta\theta}{\dot{\theta}} \frac{c}{\lambda 2pk^2},$$

or

$$N = \frac{f}{2pk^2}\left(\frac{\Delta\theta}{\dot{\theta}}\right), \tag{4.26}$$

where f = frequency in Hz.

A convenient form for equation 4.26 is

$$2pN = \left(\frac{f}{k^2}\right)\left(\frac{\Delta\theta}{\dot{\theta}}\right), \tag{4.27}$$

which is plotted for a range of values in figure 4–23.

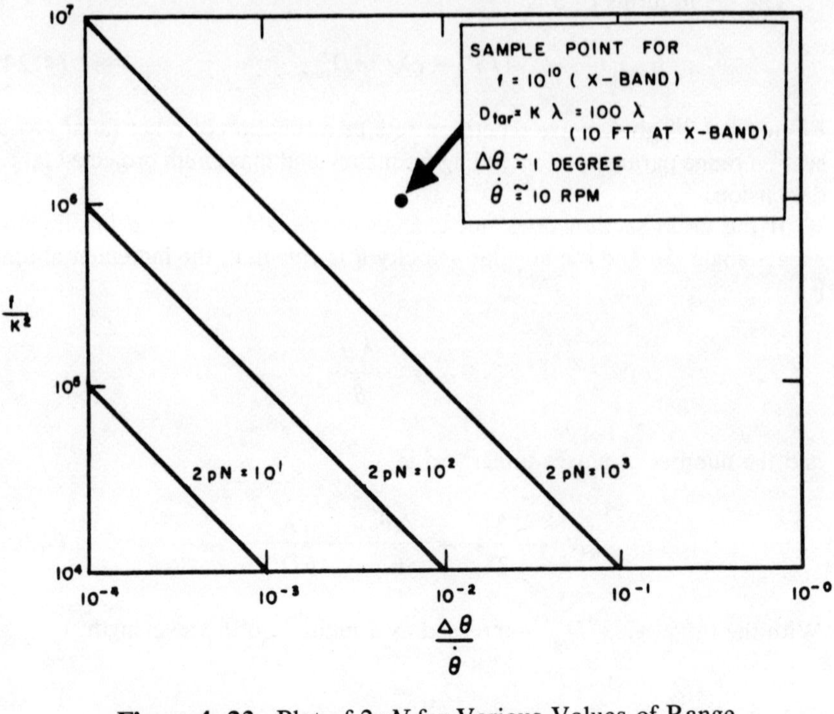

Figure 4–23. Plot of $2pN$ for Various Values of Range

Scale Measurements of Radar Cross-Section

The use of scaled target models and frequencies is justified by the principle of electrodynamic similitude—a direct consequence of the linearity of Maxwell's equations [45,46]. Models have been successfully used to study the properties of antennas for many years [53]. For radar cross-section measurements, the question arises: Under what conditions is possible to use scale models?

The conditions for proper scale metal-model measurements requires that the conductivity of the model be scaled accordingly [54]. However, in measurement of cross-section of highly conducting targets, the conductivity is rarely scaled. Models of high conductivity have only a small tangential component of electric-field intensity at the surface; and the difference between the surface current distribution on a poor conductor and the current distribution on a perfectly conducting target is very slight.

Some of the linear scaling principles can be explained by a few mathematical terms, as follows:

The scaling equations [46], as shown from Maxwell's equations, are

$$\mu\varepsilon l^2 f^2 = C_1, \tag{4.28}$$

and

$$\mu\sigma l^2 f = C_2, \tag{4.29}$$

where μ = permeability of the medium,

ε = permittivity of the medium,

l = linear dimension of the target and/or model,

f = frequency, and

σ = conductivity.

The condition of similitude requires that the two characteristic parameters C_1 and C_2 be invariant to a change of scale.

In the usual high-conducting target such as steel, brass, or copper, the currents in the conductors are essentially the same as in ideal conductors [48]. Therefore, the target conductivity can be assumed to be infinite; thus it is only necessary to consider equation 4.28,

$$\mu_m \varepsilon_m l_m^2 f_m^2 = \mu\varepsilon l^2 f^2, \tag{4.30}$$

where the subscript m represents the parameters of the model. In the usual case, $\mu_m = \mu$, and $\varepsilon_m = \varepsilon$; therefore, equation 4.30 is

$$\frac{lm^2}{l^2} = \frac{f^2}{f_m^{\,2}}, \tag{4.31}$$

which shows that the dimensions of the model scale linearly with frequency.

When considering lossy dielectric-coated targets, the scaling relation using equation 4.29 is

$$\mu_m \varepsilon_m l_m^{\,2} f_m^{\,2} \tan \delta_m = \mu \varepsilon l^2 f^2 \tan \delta, \tag{4.32}$$

where tan δ represents the loss tangent of the dielectric material (the loss tangent is related by $\tan \delta = \sigma/\omega\varepsilon$). If the parameters of the modeled dielectric are the same as those of the full-scale dielectric, then the linear relations of equation 4.31 are maintained. This requires that the dielectric material of the model be the same as that of the full-scale target. Permittivity of common dielectric materials such as Teflon, paraffin, and nylon are practically independent of frequency below 34 GHz.

Targets are often coated with ablating materials, lossy dielectrics, and ferrites and may also be surrounded by a plasma sheath. Ablating and absorbing materials in general do not have permeability and permittivity independent of frequency. Under these conditions scaling may not be possible. Linear modeling is very limited in extent for nonconducting targets. The University of Michigan has demonstrated the feasibility of the nonlinear scaling theory [47]. However, the nonlinear scaling theory is applicable to only a few specialized practical problems.

Specialized radar cross-section measurements have been developed using wide-bandwidth transmitted waveforms. Such waveforms allow individual scattering centers on a radar target to be located. Once known, these can be minimized to reduce the overall radar cross-section or selectively enhanced to alter it. Short-pulse, linear-frequency modulation and pseudorandom phase-coded modulated waveforms have been investigated for this purpose [55].

Quick-Check Tests for Cross-Section Ranges[12]

In determining the suitability of a cross-section range for measurement purposes, a knowledge of system characteristics and errors is essential. There are many intangibles such as operator skills that are difficult to evaluate in a short series of tests. However, a selected series of tests can give the desired knowledge of system characteristics and errors. Although it is possible to devise elaborate and lengthy checkout procedures, a quick indication of system capabilities can be achieved with the following series of tests:

1. system-linearity test,
2. system-stability test,
3. far-zone test,
4. background-contribution test,
5. target-support-contribution test,
6. forward-scatter-modification test,
7. cross-coupling between target and ground-plane image test.

System-Linearity Tests

A check of system linearity can be made by measuring the return of a set of calibration spheres and other targets such as corner reflectors, if necessary, to show the limits of the linear operation range of the overall system.

System-Stability Tests

A constant signal is introduced into the RF portion of the system, and system stability is checked at the various levels at which the system is expected to operate. A plot of system output versus time is made for the various constant signal levels introduced.

Far-Zone Tests

Far-zone conditions are checked by making measurements at various ranges with an object such as a cone to show consistency of data as range is changed.

Background-Contribution Tests

Since the background contribution can be a source of considerable error in the measurement of small cross-sections, several tests are required. Plots of each of the calibration spheres are made at several ranges with an incremental range change of one-quarter wavelength between successive measurements of the same sphere. Plots of each of the calibration spheres set off center by one-quarter wavelength on the supporting column are also made. In both tests, plots of the system background are made at the same system gain setting.

Target-Support-Contribution Tests

Although the target-support contribution is possibly a part of the interfering background level, a separate test is required to determine the contribution of the support column or support device to be used in supporting full-scale vehicles. A plot is required of the interfering background level in the absence of the support column with suitable calibration with several spheres. Without change in the system, the support column is then introduced and a plot made of the system background level as the column is rotated through a full 360°. As a final test of the support column, a continuous nose rock of the support column over a full one-wavelength interval is recorded.

Forward-Scatter Modification

A serious difficulty encountered in measuring full-scale vehicles is modification of forward scatter by the target. A flat plate, approximately 2 ft square, is measured with the plate tilted at a 45° angle to the incident beam. A comparison of front-to-back response of the flat plate indicates difficulty with forward-scatter modification.

Cross-Coupling between Target and Ground-Plane Image

The presence of the ground in cross-section measuring can contribute errors in measuring certain objects. Cross coupling between the target and its image in the ground plane is obtained by plotting the response of the flat plate (used for the previous tests) tilted at various angles with respect to the incident beam. Ground-plane effects are indicated by difference in front-to-back response of the plate at certain tilt angles.

Notes

1. The radar cross-section of birds may be as high as 0.01 m^2 at L-band. See R.E. Richardson, J.M. Stacey, H.M. Kohler, and F.R. Naka, "Radar Returns from Birds and their Elimination from Radar Outputs,"

Lincoln Laboratory, Group Report 45-42, AD 241994, 22 December 1959. The radar cross-section of a June bug is 10^{-5} m^2 at X-band. See Cornell Aeronautical Laboratory advertisement, *Proc. IRE* 50(April 1962): 94A.

2. Throughout this chapter the following definitions are used: dBm^2 refers to a decibel with respect to 1 square meter (that is, $dBm^2 = 10 \log \sigma$); $dB\lambda^2$ refers to a decibel with respect to one σ/λ^2 (that is, $dB\lambda^2 = 10 \log(\sigma/\lambda^2)$).

3. The IEEE Antennas and Waveguides Committee has decided to define the *near field* as containing a radiating near-field region and a reactive near-field region immediately surrounding the aperture.

4. Absorbers can be tested individually and selected for reflection coefficients as low as −60 dB. However, with aging and soiling no guarantee is available that the low reflection coefficient will be maintained. Styrofoam columns can also be selected, which have somewhat lower reflection coefficient.

5. Private communication.

6. This is the factor by which the beamwidth is broadened and by which the gain of a line source is reduced.

7. Uncolored nylon rope is to be preferred, as coloring dyes frequently contain metallic salts that noticeably increase radar cross-section.

8. Alternatively, a short-slot hybrid coupler or similar device may be used in place of the magic T.

9. It is unfortunate that common usage includes σ for both standard deviation and scattering cross-section.

10. An interesting solution to the problem of wideband amplifiers with a wide dynamic range is the logarithmic traveling tube [35].

11. Courtesy of the Conductron Corporation.

12. Courtesy of A.W. Wren, Jr.

References

1. Cumming, W.A. "Radiation Measurements at Radio Frequencies: A Survey of Current Techniques." *Proc. IRE* 47(May 1959): 705–735.
2. Silver, S. *Microwave Antenna Theory and Design.* New York: McGraw-Hill, 1949.
3. Bickmore, R.W., and Hansen, R.C. "Antenna Power Densities in the Fresnel Region." *Proc. IRE* (*Correspondence*) 47(December 1959): 2119–2120.
4. Hansen, R.C., and Bailin, L.L. "A New Method of Near Field Analysis." *Trans. IRE*, AP-7, Special Supplement (December 1959): S458–S467.
5. Brown, J. "A Theoretical Analysis of Some Errors in Aerial Measure-

ments." *Proc. Inst. Elec. Eng.* (London) 105, pt. C (September 1958): 343–351.
6. Bickmore, R.W. "On Focusing Electromagnetic Radiators." *Can. J. Phys.* 35(1957): 1292–1298.
7. Dunbar, A.S. "Applications of the Rayleigh Criterion, Part I." Report AFCRC-TR-59-118(I), McGill University Symposium on Microwave Optics, Montreal, Canada, April 1959, pp. 63–66.
8. Bickmore, R.W. "Fraunhofer Pattern Measurement in the Fresnel Region." *Can. J. Phys.* 35(1957): 1299–1308.
9. Shinn, D.H. "Misfocusing and the Near-Field of Microwave Aerials." *Marconi Rev.* 19, no. 123 (1956): 141–149.
10. Rhodes, D.R. "On Minimum Range for Radiation Patterns." *Proc. IRE* 42(September 1954): 1408–1410.
11. King, R.W.P., and Wu, T.T. *The Scattering and Diffraction of Waves.* Cambridge, Mass. Harvard University Press, 1959, Chap. 6.
12. Melling, W.P. "Errors in the Measurement of Radar Echoing Patterns: A New Minimum Range Criterion." Report no. UB-1088-P-100, Cornell Aeronautical Laboratory, 9 January 1957.
13. Hu, M.K. "Fresnel Region Fields of Circular Aperture Antennas." *J. Res. NBS* 65D(March–April 1961): 137–147.
14. Melling, W.P. "An Analysis of Radar Cross-Section Measurement Techniques." Report no. UB-1088-P-104, Cornell Aeronautical Laboratory, 18 September 1959.
15. Gilfert, J. "Memorandum on the Reduction of Radar Echo from Polyfoam Model Supports." Report no. 475-7, Ohio State University, 30 October 1953.
16. "The Echoing Properties of Missiles, Part II: Measurements" Presented at Royal Aircraft Establishment Symposium, Farnborough, England, 8–10 June 1960.
17. Cram, L.A.; Prior, G.M.; and Oxley, B. "A Millimetre Pulse Radar for Indoor Echoing Area Measurements" Report D.P. 736, EMI Electronics, Ltd., June 1960.
18. Burrows, C.B., and Attwood, S.S. *Radio Wave Propagation.* New York: Academic Press, 1949.
19. Jordan, E.C.. *Electromagnetic Waves and Radiating Systems.* Englewood Cliffs, N.J.: Prentice-Hall, 1950, p. 572.
20. Cohen, A., and Maltese, Q.W. "The Lincoln Laboratory Antenna Test Range." *Micro. Jour.* (April 1961): 57–65.
21. Scheldorf, M.W. "Proposed Technique for Antenna Pattern Measurement." Presented at Sixth Radio Frequency Interference Conference, Chicago, 4–6 October 1960.
22. Cutler, C.C.; King, A.P.; and Kock, W.E. "Microwave Antenna Measurements." *Proc. IRE* 35, December 1947, pp. 1462–1471.

23. Gorr, B.B. "AFCRC Radar Cross-Section Measuring Equipment and Range." Report AFCRC-TR-59-137, April 1959.
24. Honda, J.S. "Scattering of Microwaves by Figures of Revolution." IRE-WESCON Conv. Rec., pt. 1, 1957, pp. 151–157.
25. Caldecott, R., and Burgess, J. "A System for the High Speed Recording of Radar Echo Patterns and for the Transfer of the Data to Punched Tape." Report no. 792-2, Ohio State Antenna Laboratory, 15 December 1958.
26. Scharfman, H., and King, D.D. "Antenna-Scattering Measurements by Modulation of the Scatterer." *Proc. IRE* 42(May 1954): 854–858.
27. McMillan, E.B., and Schmitt, H.J. "Doppler Method for Absorber Testing." *Micro. Jour.* 3(November 1960): 64–68.
28. Schmitt, H.J. "Back-Scattering Measurements with a Space-Separation Method." *Trans. IRE* AP-7 (January 1959): 15–22.
29. Steinbach, K., and Varnum, F.B. "A Scattering Measurement Technique." *Trans. IRE* (*Communications*) AP-9 (May 1961): 312–313.
30. Olte, A., and Silver, S. "New Results in Backscattering from Cones and Spheroids." *Trans. IRE* AP-7 (December 1959): S61–S67.
31. Bacon, J., and Burgess, J. "A Time Gate for Echo-Measuring Radar Installations." Report no. 792–4, Ohio State Antenna Laboratory, 15 January 1959 (ASTIA no. AD 212351).
32. Gillespie, N.R., et al. "The Evolution and Application of Coherent Radar Systems." *Trans. IRE* MIL-5 (April 1961): 131–138.
33. Ridenour, L.N. *Radar System Engineering*. MIT Radiation Laboratory Series, vol 1. New York: McGraw-Hill, 1947, p. 143.
34. Sabih, D. "A Method to Reduce Antenna Ground Reflections." *Trans. IRE* (*Communications*) AP-8 (March 1960): 225–227.
35. Kliger, J.; Downey, E.J.; and Munushian, J. "Improvement of R-F Dynamic Range." Report RADC-TR-59-222, Hughes, 28 December 1959.
36. Simmons, A.J. "Theoretical Estimate of Background Backscattering Cross Section of the Lincoln Laboratory Antenna Range." Report no. 221-1, Technical Research Group, 28 Feburary 1962.
37. Buckley, E.F. "The Design and Evaluation of Microwave Anechoic Chambers." Internal memo, Emerson and Cumming, Inc., April 1960.
38. Crandell, P.A. "A Turnstile Polarizer for Rain Cancellation." *Trans. IRE* MTT-3 (January 1955): 10–15.
39. Potter, R.S. "The Analysis and Matching of a Trimode Turnstile Waveguide Junction." Report 4670, Naval Research Laboratory, 19 December 1955.
40. Allen, P.J., and Tompkins, R.D. "An Instantaneous Microwave

Polarimeter Technique." Report 5214, Naval Research Laboratory, 29 September 1958.

41. "Measurement of the Radar Cross Section of a F4H at Ka-Band." Technical Report no. 0230, Sierra Research Corporation, Buffalo, N.Y., January 1962.
42. Osofsky, I.B. Private communication to C.G. Bachman, Cornell Aeronautical Laboratory, April 1962.
43. Garbacz, R.J. "The Bistatic Scattering from a Class of Lossy Dielectric Spheres with Surface Impedance Boundary Conditions." Report no. 925-5, Ohio State University Antenna Laboratory, 15 January 1961.
44. Adler, S.B. "Pulsed Radar Measurement of Back Scattering from Spheres." *RCA Rev.* 23, no. 1 (March 1962): 80–95.
45. Stratton, J.A. *Electromagnetic Theory*. New York: McGraw-Hill, 1941, pp. 488–490.
46. King, R. *Electromagnetic Engineering*, vol. 1. New York: McGraw-Hill, 1945, pp. 316–320.
47. Belyrea, J.E., et al. "Studies in Radar Cross Sections XLV, Studies in Non-Linear Modeling-II." Final Report no. 2871-6-F, AFCRL-40, University of Michigan, Radiation Laboratory, 31 December 1960.
48. Ramo, S., and Whinnery, J.R. *Fields and Waves in Modern Radio*, 2nd ed. New York: Wiley, 1953, p. 323.
49. Harrington, R.F. *Time-Harmonic Electromagnetic Fields*. New York: McGraw-Hill, 1961, pp. 292–296.
50. Keys, J.E., and Primich, R.I. "The Experimental Determination of the Far-Field Scattering from Simple Shapes." *Trans IRE* AP-7 (Special Supplement): S77. Work performed at the Defense Research Telecommunications Establishment, Canada, December 1959.
51. Kell, R.E., and Hendrick, L. "Re-examination of Minimum Range Criteria for Radar Cross Section Measurements." Report no. GM 1580-G-101, Cornell Aeronautical Laboratory, 3 August 1962.
52. Newton, D.J. "The Delta Radar System, A Wide Bandwidth FM/CW X-Band Radar for the Fine Range Resolution Measurement of Radar Cross Section." Final Report no. UB-1571-P-1, Cornell Aeronautical Laboratory, 31 July 1962.
53. Sinclair, G.; Jordan, E.C.; and Vaughn, E.W. "Measurement of Aircraft-Antenna Patterns Using Models." *Proc. IRE* 35, no. 12 (December 1947): 1451–1462.
54. Mentzer, J.R. *Scattering and Diffraction of Radio Waves.* New York: Pergamon Press, 1955, chap. 5.
55. Bachman, C.G. "Some Recent Developments in RCS Measurement Techniques." *Proc. IEEE* 53, no. 8(August 1965): 962–972.

5 Radar-Target Identification

Radar measurements can be used to determine a target's properties of size, shape, and motion by extraction of information from the return signal. In general, such signal processing for target identification exploits the differences between transmitted and received waveforms in terms of their relative amplitudes, phases, and polarizations. The time histories of these return signals, or certain of their properties as obtained by filtering, are compared with data resident in a reference file to effect target identification. Other information may be provided to improve the identification process [11]. The effects of propagation anomalies must be known to compensate for the signal variations along the propagation path from transmitter to reflecting target to receiver, or multiple receivers, as is the case in some radar systems. The geometry relating transmitter locations, target locations, and receiver locations must be known also in some target-recognition schemes. Many radar-system requirements may be to isolate a particular target in a cluster or group of targets resolvable in range and angle. In such an operation the target-identification process is simply to select or discriminate one target from the others.

Identifier subsystems may be implemented by techniques ranging from amplitude thresholding, or velocity filtering, to the use of complex signal-processing algorithms in pattern-recognition schemes requiring large-capacity digital computers and extensive software. Figure 5–1 illustrates a general identification process for aircraft to determine whether the radar contact is a friend, a foe, or a neutral, by passive methods in that the radar target does not carry an identifying beacon. All target-scattering properties are of possible utility in such identification schemes [5,10].

A fundamental target-scattering property is its amplitude return as a function of frequency, as shown in figure 5–2. In the low frequency, or Rayleigh region, radar cross-section amplitude is proportional to the target volume, whereas in the resonance region, where target dimensions are comparable to nominal wavelength, amplitude oscillations are observed [2,3,6]. The high-frequency, or optics region, where target dimensions are large compared with wavelength, requires decomposition of a complex target into elemental scattering surfaces for target identification [6,8]. A wide-bandwidth signal of either a pulse duration short in equivalent-range extent to the maximum target dimension along the line of sight, or a ramp amplitude time waveform, may be used [4,6,9,10].

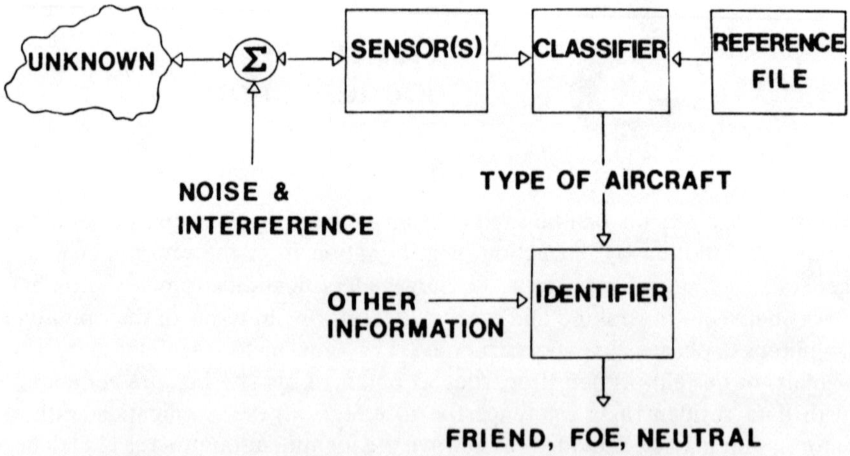

Figure 5–1. General Concept of the Identification Process

The low-frequency target-identification method may be implemented by the system archetecture outlined in figure 5–3. Amplitude and phase measurements of the time-frequency ramp return signal forms the basis for construction of a three-dimensional target image by extensive signal processing [3].

One wide-bandwidth identifier scheme is shown in figure 5–4. This method uses range time and relative doppler of point-scattering centers on

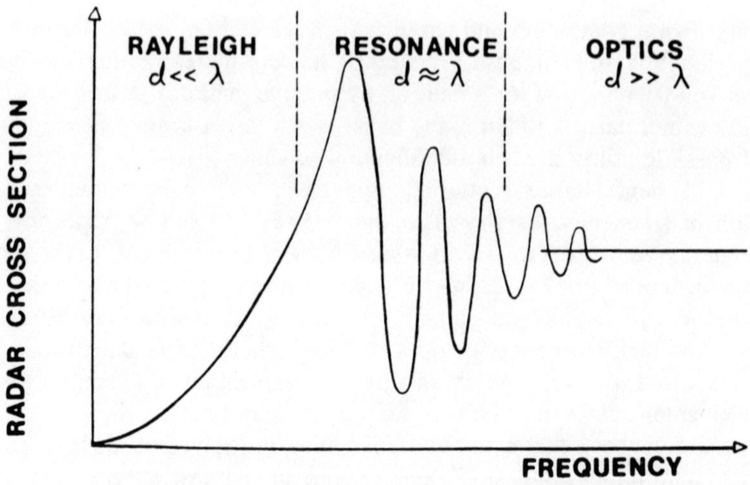

Figure 5–2. Frequency-Domain Characteristics of Radar-Target Scattering

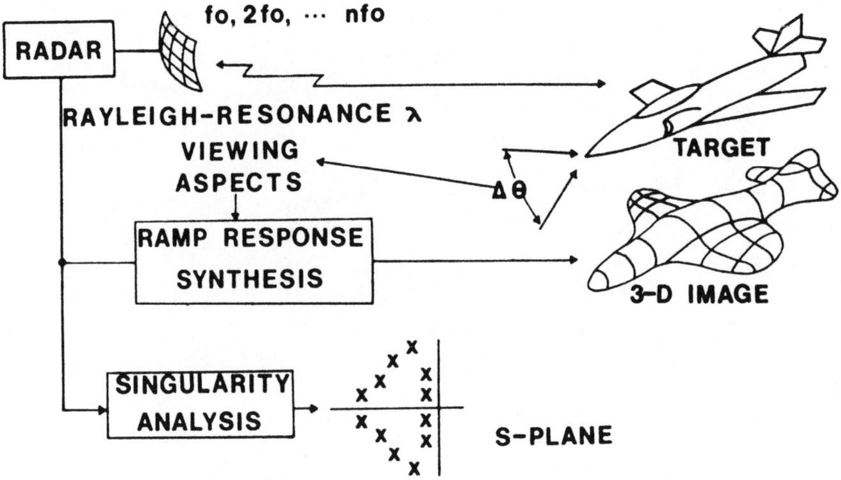

Figure 5–3. Low-Frequency Radar Identifier

the target to allow construction of a two-dimensional target image. Amplitude time signals of inline, or slant range returns, yields one dimension; relative doppler frequency amplitudes provide cross-range dimensions to complete the target portrait. Extensive computer memory and data processing are required for this type of image reconstruction, with data taken over relatively long times and viewing aspect angles [4].

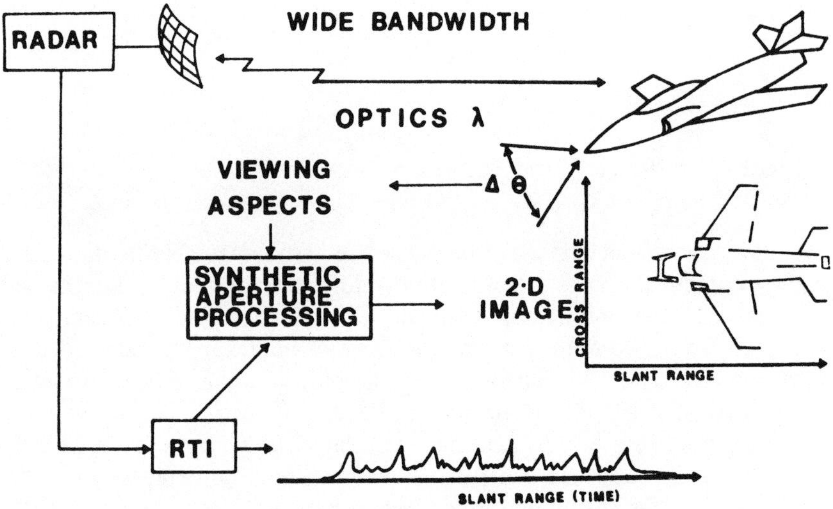

Figure 5–4. Microwave Radar Identifier

Rate	*Target Property*
$\left.\frac{\partial A}{\partial t}\right\|_{f,R \text{ const.}}$	→ Motion
$\left.\frac{\partial A}{\partial f}\right\|_{t,R \text{ const.}}$	→ Size
$\left.\frac{\partial A}{\partial R}\right\|_{t,f \text{ const.}}$	→ Shape
$\left.\frac{\partial \psi}{\partial t}\right\|_{f,R \text{ const.}}$	→ Motion
$\left.\frac{\partial \psi}{\partial f}\right\|_{t,R \text{ const.}}$	→ Size
$\left.\frac{\partial \psi}{\partial R}\right\|_{t,f \text{ const.}}$	→ Shape

A = amplitude
ψ = phase
f = frequency
t = time
R = range

Figure 5–5. Radar-Target Identifiers

Measurements of signal returns are useful in target identification because they contain derivable identification information based on the changes in signal amplitude, phase, and frequency. Polarization is also valuable but not as well developed as other techniques. For example, an extended edge target in one dimension will have different-amplitude signal returns for incident fields parallel to and orthogonal to the edge. Comparison of these orthogonally polarized returns can indicate relative target dimensions [8,10].

The information content of signal return may be expressed in a set of rates of changes in amplitude and phase versus time, frequency, and distance.

Figure 5–5 illustrates the mathematical expressions of these rates as partial derivatives and the type of target information each can provide [11].

Different basic radar-target shapes form sections of a complex scattering configuration. The variations of signal amplitudes as a function of operating wavelength are useful in determining the radar cross-sections of the elemental bodies over the low, resonance, and geometric-optics scattering regions. A convenient plot of these variations, normalized to wavelengths, is given in figure 5–6 for the geometrical configurations of spheres, cylinders, and cone-spheres. This may be used to estimate the radar cross-section of radar targets having similar individual or composite shapes for given wavelengths. Figure 5–7 indicates the sensitivity to frequency of peak specular broadside radar cross-section levels for three generic aerospace-vehicle fuselage shapes.

Radar cross-section magnitude prediction formulas have been derived, using the mathematical foundations of geometric optics [5,8,11]. These are listed in figure 5–8 for a number of elemental scatterers that are components of more complex radar targets. These may be used to estimate radar target cross-sections by computing each value at a particular aspect angle and takin the square root of the sum of the squares of these values. Accuracy obtained

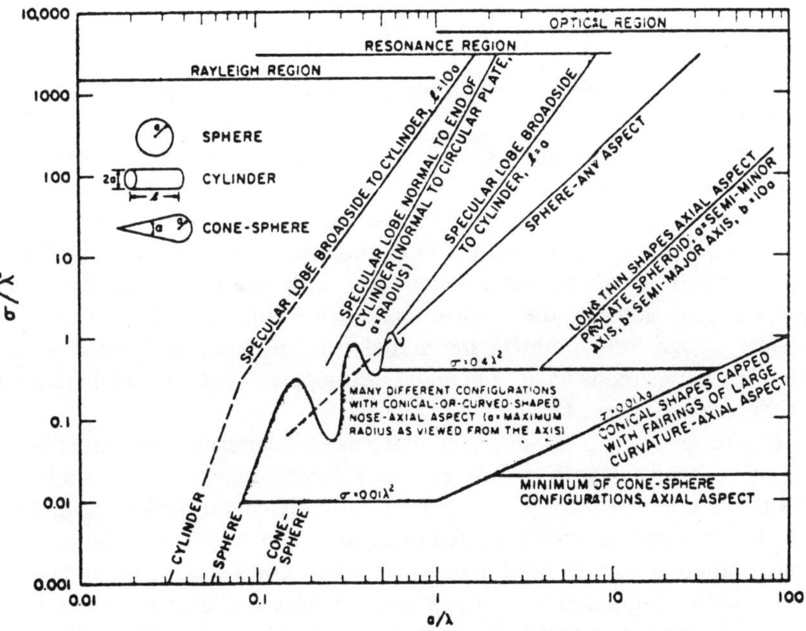

Figure 5–6. Backscattering RCS of Possible Configurations

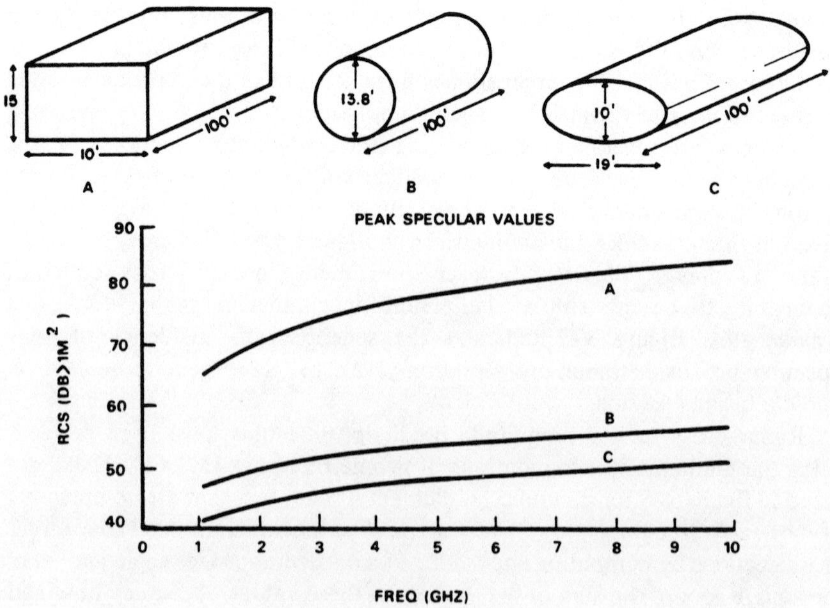

Figure 5–7. Effect of Fuselage Shape on Broadside RCS

in this way can be within ±6 dB of the measured value of radar cross-section [6,8].

Symmetrical scattering bodies will have a radar cross-section amplitude that varies with aspect angle. This fact forms the basis of many target-identification schemes where orbiting satellites passing overhead are viewed from ground-based radars and the amplitude-signal time histories recorded. The complex amplitude-data record is analyzed to extract elemental composite portions of the vehical and then, with knowledge of general satellite design, to synthesize the actual configuration. Such space-object identification methods have been under development and refinement for over two decades [6,7,10,11].

Figures 5–9 and 5–10 illustrate patterns for scattering from a number of scattering bodies, showing how to relate target dimensions to amplitude versus angle lobe widths. The combination of elemental bodies into generic vehicle configuration are illustrated in figures 5–11 and 5–12. The rounded configuration is seen to have a smoother radar cross-section variation with aspect angle, compared with the example with sharp discontinuities in its shape. The maximum length of a radar target may be estimated, based on the

Scatter	Aspect	Radar Cross Section	Definitions Symbols
Curved Surface	Radial	$\pi R_1 R_2$	$R_1, R_2 =$ Principal radii of curvature
Sphere	All	πa^2	$a =$ Radius
Finite Cone	Axial	$\pi a^2 \tan^2 a/2$	$a =$ Base Radius $a =$ Cone Angle
Semi-Infinite	Axial	$\dfrac{\lambda^2 \tan^4 a/2}{16\pi}$	$a =$ Cone Angle
Prolate Spheroid	Axial	$\pi a^4/b^2$	$a =$ Semiminor axis $b =$ Semimajor axis
Ogive	Axial	$\dfrac{\lambda^2 \tan^4 \gamma/2}{16\,\pi}$	$\gamma =$ Angle of tangent nose cone
Paraboloid	Axial	$4\pi\,\rho^2$	$\rho =$ Apex radius of curvature
Flat Surface	Normal	$4\pi A^2/\lambda^2$	$A =$ Surface Area
Circular Plate	Incidence at angle θ to normal	$\pi a^2 \,\mathrm{ctn}^2\, \theta l_1 \left(\dfrac{4\pi a}{\lambda}\right) \sin\theta$	$a =$ Radius of plate
Circular Cylinder	Incidence at angle θ to broadside, θ measured in axial plane	$2\pi\, al \cos\theta \left[\dfrac{\sin^2(2\pi\, l \sin\theta/\lambda)}{(2\pi l \sin\theta)^2}\right]$	$a =$ Radius $t =$ Cylinder length

Figure 5–8. Formulas for the Backscattering Cross-Sections of Large Characteristic Dimensions

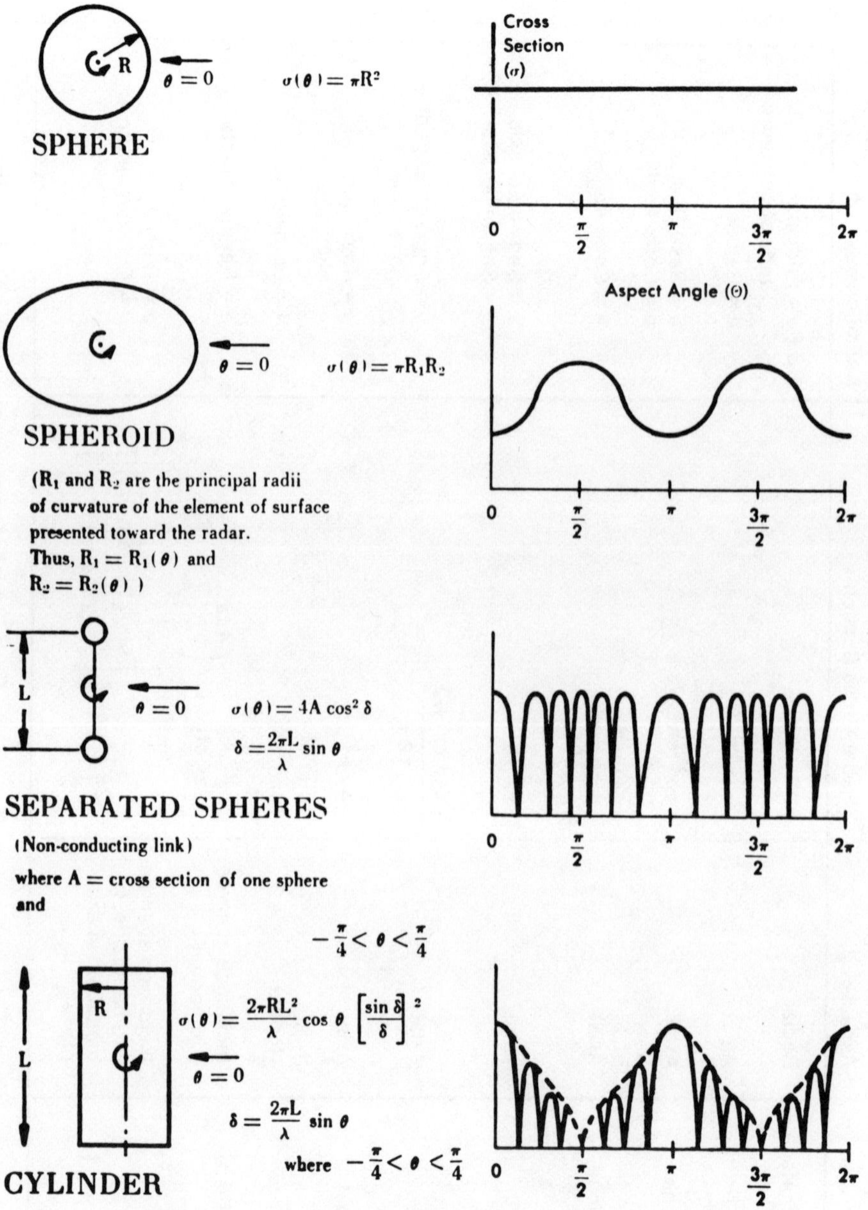

Figure 5–9. Typical Patterns for Symmetrical Bodies

CYLINDER

$$\sigma(\theta) = \frac{2\pi R L^2}{\lambda}\cos\theta\left[\frac{\sin\frac{2\pi L}{\lambda}\sin\theta}{\frac{2\pi L}{\lambda}\sin\theta}\right]^2$$

$$L = \frac{\lambda}{2\Delta}$$

RECTANGULAR PLATE

$$\sigma(\theta) = \frac{4\pi A^2}{\lambda^2}\cos^2\theta\left[\frac{\sin\frac{2\pi x}{\lambda}\sin\theta}{\frac{2\pi x}{\lambda}\sin\theta}\right]^2$$

where $A = ab \quad x = a, b$

ROUND PLATE

$$\sigma(\theta) = \frac{4\pi A^2}{\lambda^2}\cos^2\theta\left[\frac{2J_1\frac{2\pi R}{\lambda}\sin\theta}{\frac{2\pi R}{\lambda}\sin\theta}\right]^2$$

where $A = \pi R^2$ and
J_1 = First Order Bessel Function

Note: Above three expressions hold for $-\frac{\pi}{4} < \theta < \frac{\pi}{4}$

CONE

No rigorous closed form expression known for $\sigma(\theta)$

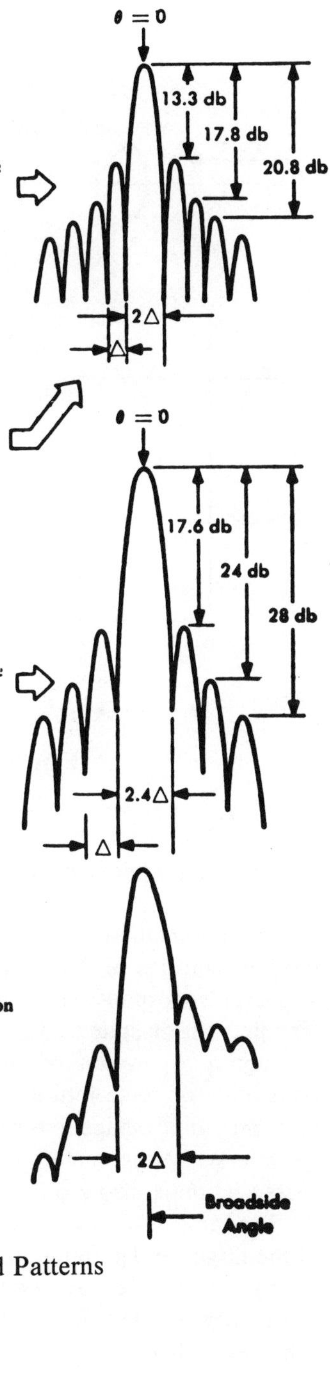

Figure 5–10. Detailed Patterns

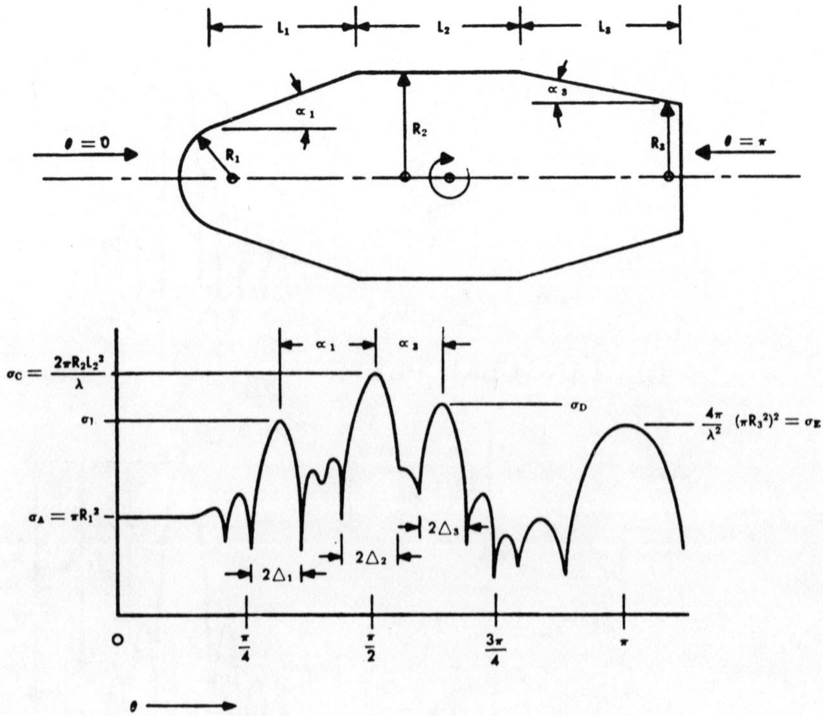

Figure 5–11. Compound Body

minimum angle between nulls in the pattern, allowing target identification by its length.

Another method of target identification by its length is to use impulse and ramp waveforms to determine target geometrics, as shown in figure 5–13. The impulse response resulting from partial returns along the scattering body pinpoints the discrete scattering centers in range along the radar line of sight. Maximum separation of such signals yields a length estimate directly. Amplitude of these returns between endpoints profiles the target in effect, thus providing shape information. Use of impulse waveforms to locate discrete scattering centers on a target body for treatment to control radar cross-section is also a powerful diagnostic technique [4,6,8].

The ramp-response waveform provides an amplitude versus time (range) of the target to determine not only length, but also shape, directly from the return signal. The backscattered ramp-waveforms phase is used in some target signal-processing methods that utilize various pattern recognition schemes [7,10].

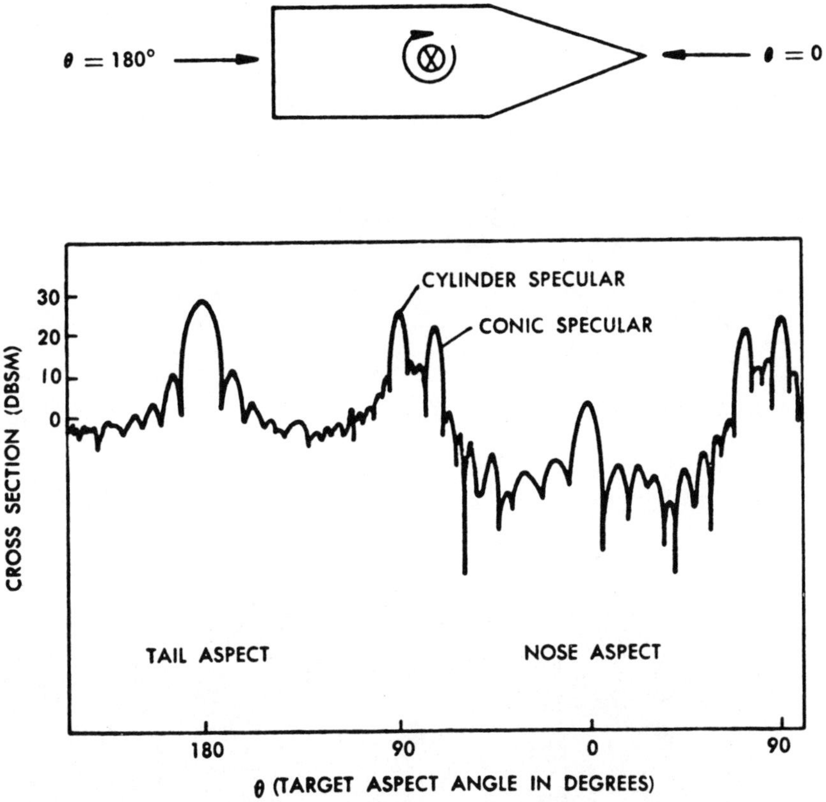

Figure 5–12. Plot of Signal Amplitude as a Function of Body Rotation Angle

Examples of actual backscattered radar signals when using impulse envelope transmitted pulse waveforms are shown in figure 5–14 for several generic targets. A sampled-data oscilloscope was used to display the return-signal amplitude versus time as shown. The point-target physical separation distance compares with the separation in range propagation time as shown on the display time base. These transmitted waveforms provided signal returns from complex large vehicles at different viewed aspect angles, as shown in figures 5–15(a), (b), and (c). In these it is possible to correlate impulse amplitudes in time with distinct parts of the vehicle in range, allowing target recognition by visual pattern recognition. A large vehicle can readily be determined when compared with signal returns from a small vehicle using such displays [8].

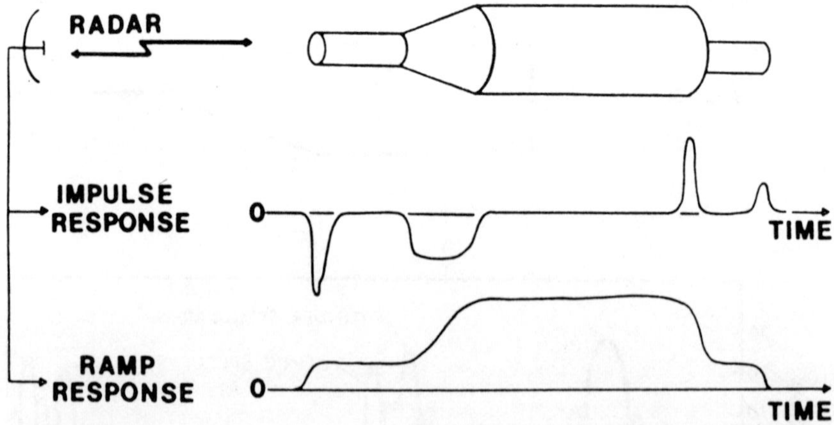

Figure 5–13. Time-Domain Characteristics of Radar-Target Scattering

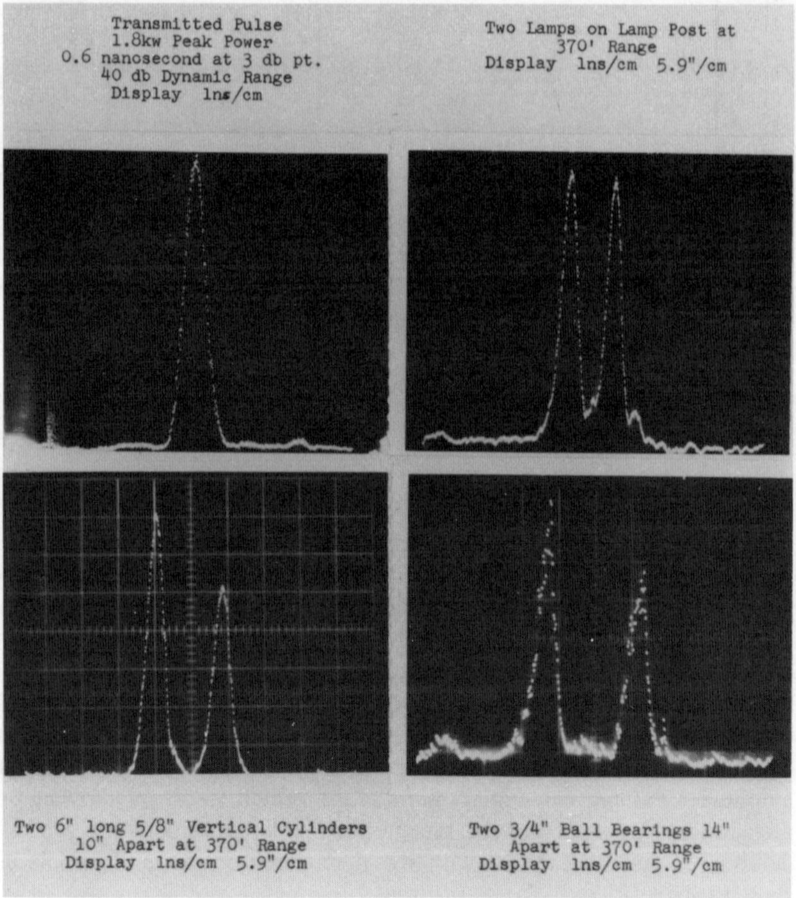

Figure 5–14. Short-Pulse Returns from Generic Radar Targets

The backscatter return amplitude-signal fine structure, as a function of viewed aspect angle, will be determined by the separation of discrete scattering centers on the target, as measured in wavelengths [12]. For fixed-target geometrics, high-frequency returns exhibit rapid oscillations, whereas low-frequency returns versus aspect angle are relatively smooth. Examples of typical aircraft echo patterns for high and low frequencies are shown in figures 5–16(a) and (b), respectively. Smoothed signal returns also occur when a great number of scattering points are present for a fixed frequency, and for a swept-frequency waveform [4,8,9].

Wide-frequency-range waveforms will allow determination of target resonances, leading to identification by comparing return signal as a function of frequency. A conically shaped radar target of fixed length has a radar cross-section versus frequency such as illustrated in figure 5–17. The peak radar cross-section value is at approximately 150 Mhz, corresponding to a wavelength of 2 m or 120 in. The length of the target yielding this data would be estimated as approximately 10 ft. A set of such determined lengths for a

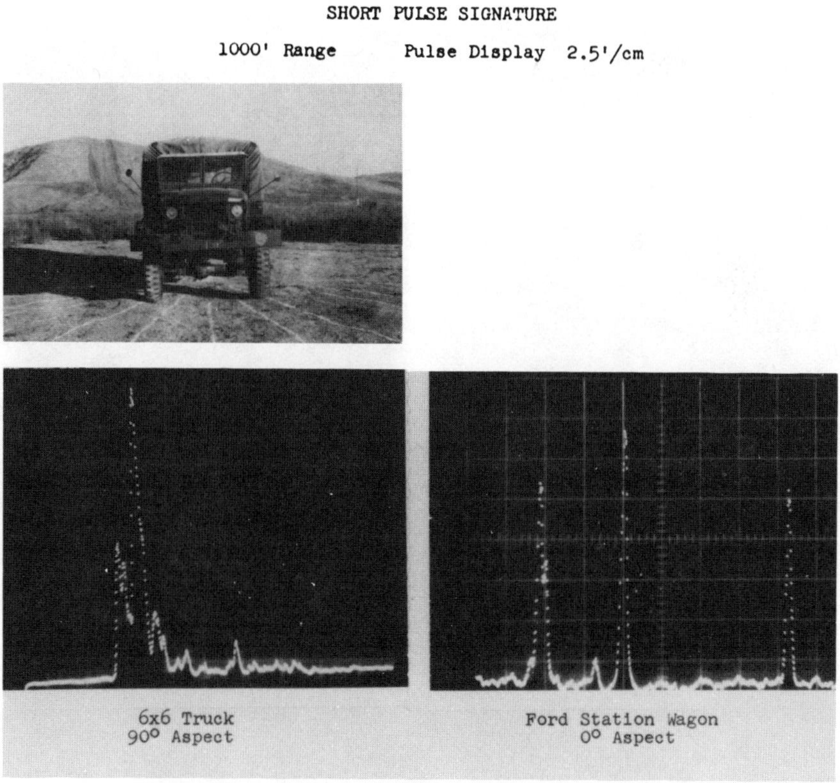

Figure 5–15(a). Vehicle Short-Pulse Returns

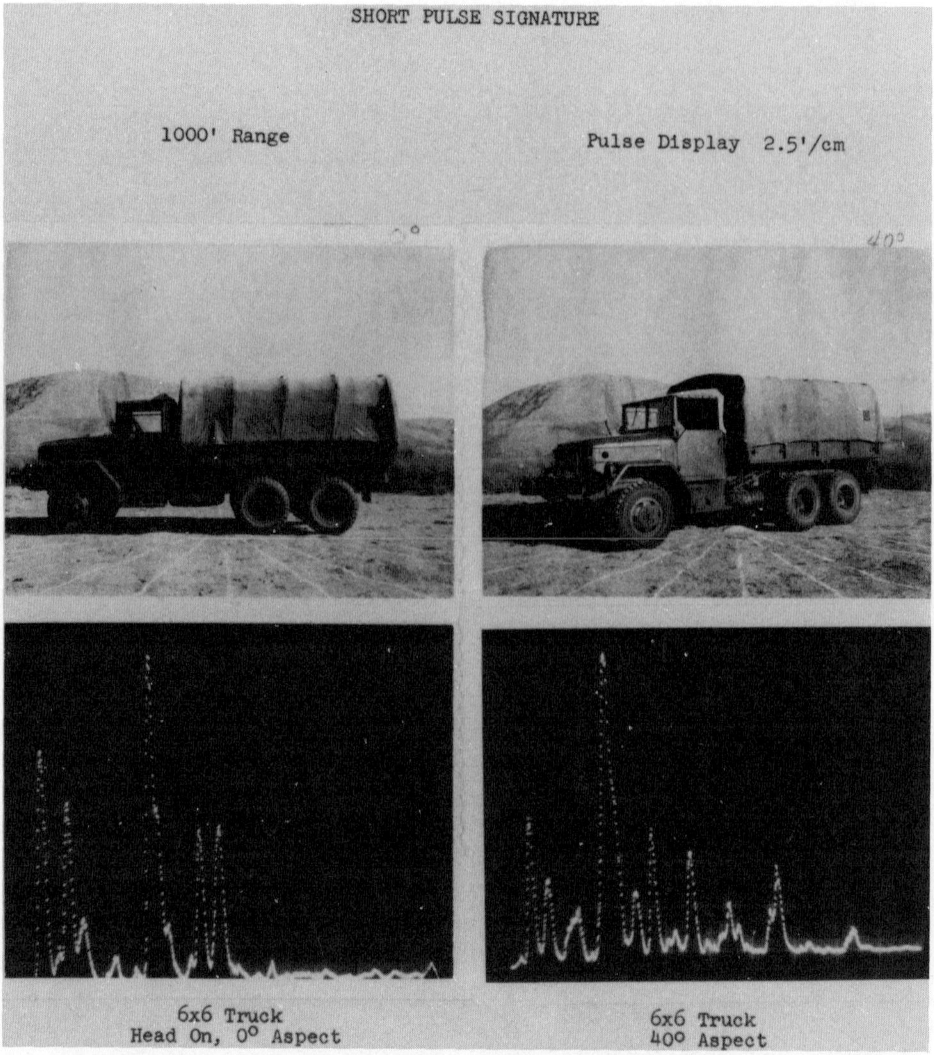

Figure 5–15(b). Vehicle Short-Pulse Returns

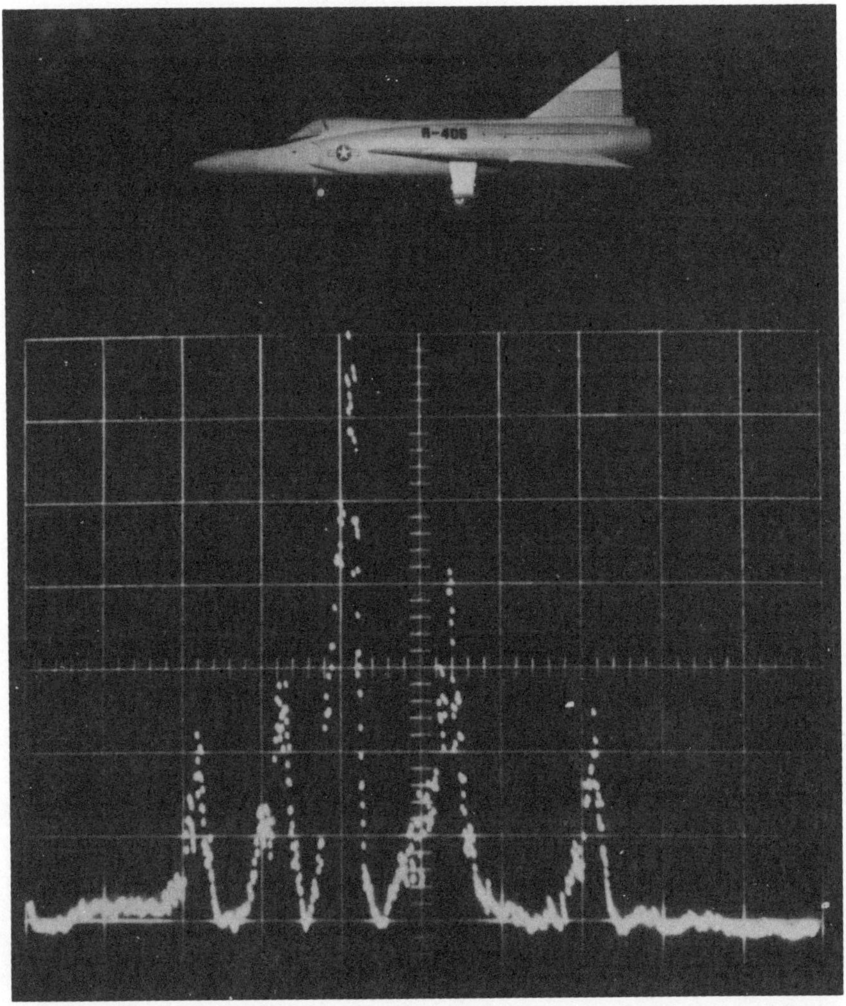

1. Battery inside radome
2–3. Structural cross-braces inside fuselage
4. Main landing-gear struts
5. Control-cable mechanism inside fuselage

Source: Courtesy of R. R. Hively, Micronetics, Inc.

Figure 5–15(c). Nanosecond Partial Returns from a Fiberglass Airplane Model Containing Metal Scatterers: From Left to Right

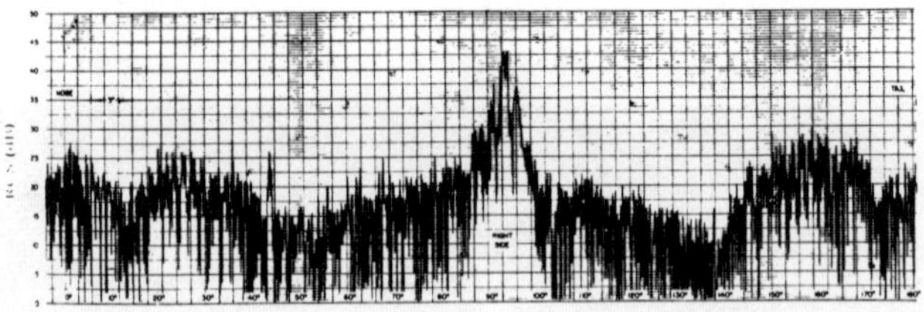

Figure 5–16(a). Typical Aircraft Echo Pattern at Microwave Frequencies

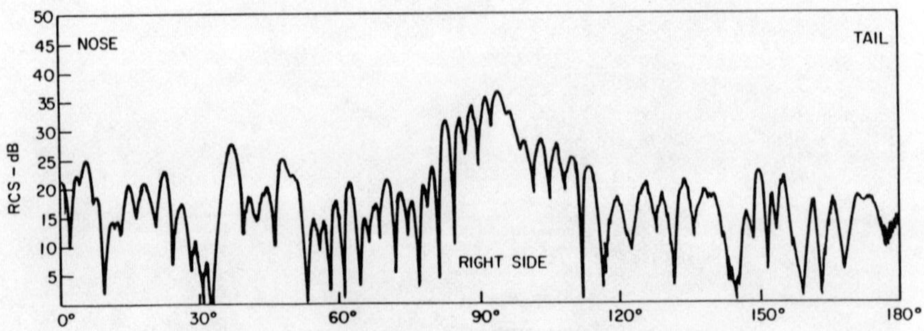

Figure 5–16(b). Long-Wavelength RCS Pattern

target in a radar range gate can be used to reconstruct the vehicle's shape [9,10].

New target-identification complex signal-processing algorithms, based on computer capability becoming available, are being developed. These radar-target-identification techniques utilize all target properties that can be sensed during an observation period and may involve monostatic, bistatic, or multistatic radars to obtain spatial interferometric data in addition to amplitude, phase, and polarization-time histories. Motional effects and geometries of reflecting objects have been determined by using mathematical target scattering matrix formulations and radar astronomy techniques.

The variations of radar cross-section amplitude for a conical body as a function of frequency and viewing angle are shown in figure 5–17(b). Each record shows radar cross-section amplitude for a discrete frequency with aspect angle, showing symmetry about nose-on. These records were arranged in order of increasing frequency, from back to front in the photo of figure 5–17(b), showing radar cross-section amplitude decrease. Radar cross-section variation with frequency is one candidate for conically shaped target identification.

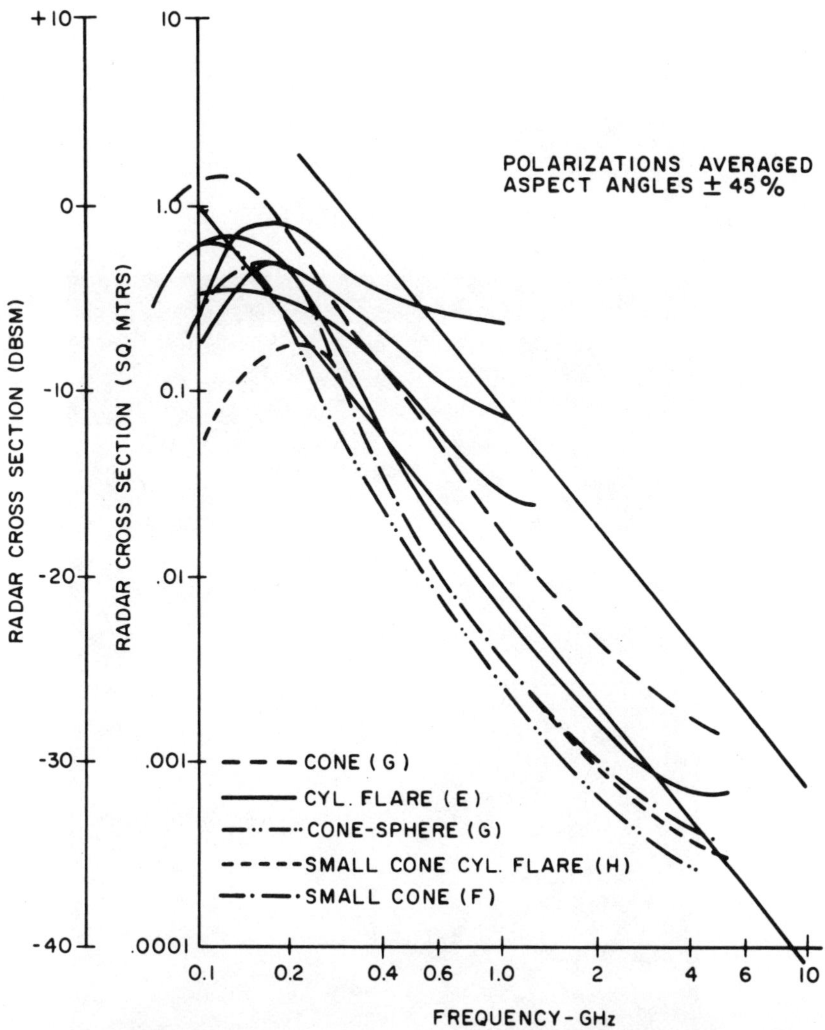

Figure 5–17(a). Radar Cross-Sections of Generic Conical Configurations

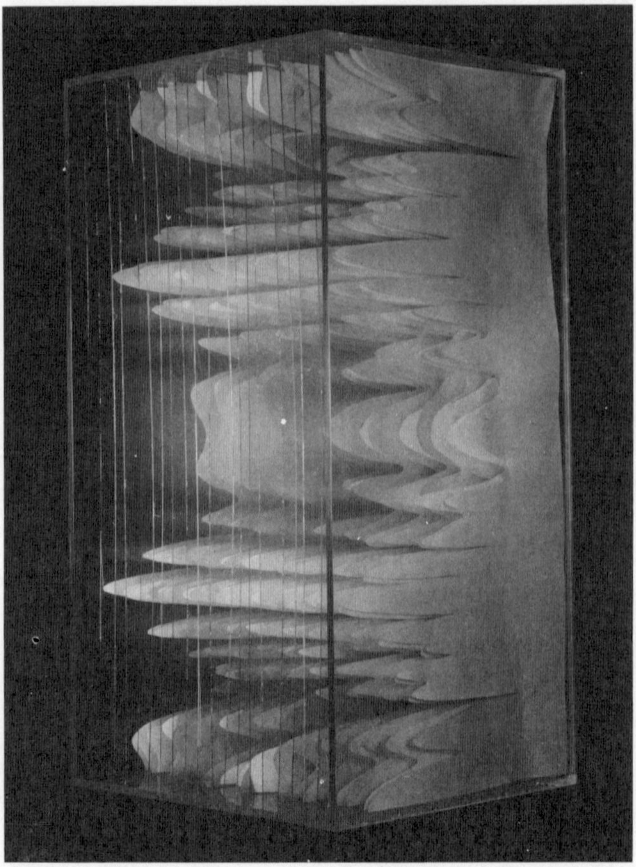

Figure 5–17(b). RCS of Cone Cylinder Flare Versus Angle and Frequency

References

1. Young, J.D. "Radar Imaging from Ramp Response Signatures." *IEEE Transactions on Antennas and Propagation* AP-24, no. 3 (May 1976): 276–282.
2. Moffat, D.L., and Mains, R.K. "Detection and Discrimination of Radar Targets." *IEEE Transactions on Antennas and Propagation* AP-23, no. 3(May 1975):358–367.
3. Ksienski, A.A.; Lin, Y-T.; and White, L.J. "Low-Frequency Approach to Target Identification." *Proceedings of the IEEE* 63, no. 12 (December 1975): 1651–1660.
4. Maceyunas, A.J., and Wallenberg, R.F. "Radar Scattering Effect from Wideband Measurements and Coherent Processing." *Proceedings of Nation Conference on Electromagnetic Scattering*, University of Illinois at Chicago Circle, 15–18 June 1976.
5. Stratton, R.D. "Real-World Limitations of Radar Signatures for Target Identification." Participant's Final Report, 1976 USAF-ASEE Summer Faculty Research Program, Rome Air Development Center, 13 August 1976.
6. "Determination and Use of Radar Scattering Characteristics." Advisory Group for Aerospace Research and Development (AGARD) Paris, September 1973 (AD-769 277).
7. *IEEE Transactions on Antennas and Propagation* (*Special Issue on Inverse Methods in Electromagnetics*) AP-29, no. 2 (March 1981).
8. *IEEE Proceedings* (*Special Issue on Radar Reflectivity*) 53, no. 8(August 1965).
9. Shubert, K.A., and Moffatt, D.L. "Swept Frequency Scattering Measurements of Aircraft." RADC-TR-79-110 (May 1979).
10. "Radar Target Identification," vols. I, II, class notes, Ohio State University, Department of Electrical Engineering, Columbus, Ohio, September 1979.
11. AGARDorgraph 40. "AVIONICS Research: Satellites and problems of Long Range Detection and Tracking." E.V.D. Glazier, E. Rechtin, and J. Voge, eds. New York: Pergamon press, 1960.
12. Bachman, C.G. "Some Recent Developments in RCS Measurement Techniques." *Proc. IEEE* 53, no. 8(August 1965): 962–972.

6 Scattering Matrix Formalism

Although it is common to speak of "the" radar cross-section of an object, it is well known that radar cross-section depends on the target shape and material; the angle (or angles, in the case of a bistatic system) at which the target is viewed; the radar frequency; and the polarization of the radar transmitting and receiving antennas. If a target is viewed at a particular angle with a single frequency, the cross-section depends on the radar polarization. The cross-section of a target for transmitting antenna polarization $\hat{p}$ and receiving antenna polarization $\hat{q}$ (assuming plane-wave illumination in free space) is given by

$$\sigma_{pq} = |\hat{q}S\hat{p}|^2, \quad (6.1)$$

where $\hat{p}$ is the unit column vector specifying polarization of the electric vector of the wave incident on the body,[1] $\hat{q}$ is the unit row vector specifying the electric vector polarization of the receiving antenna,[2] and S is the complex scattering matrix.[3] The assumptions in this formulation are:

1. The distance between receiver and body is large compared with the wavelength and with the dimensions of the scattering body.
2. The material of the scatterer is such that there are linear relationships between the field quantities at every point within the scatterer, whatever the incident field.

The scattering matrix is symmetrical ($a_{ij} = a_{ji}$) in at least two important cases:

1. bistatic scattering when the scattering body is a perfect conductor, and
2. backscattering for an arbitrary scattering body, consistent with the assumptions stated following equation 6.1

By properly choosing coordinates to take advantage of the transverse character of the field—namely, by taking a coordinate in the direction from body to receiving antenna—S becomes a 2×2 matrix and may be written in

From M.E. Bechtel, W.P. Melling, and G.A.R. Graham, *Final Report, Project ACRE*, Report UB-1405-P-1, Cornell Aeronautical Laboratory, Buffalo, May 1960.

the form:

$$S = \begin{bmatrix} a_{11}e^{i\phi_{11}} & , & a_{12}e^{i\phi_{12}} \\ a_{12}e^{i\phi_{12}} & , & a_{22}e^{i\phi_{22}} \end{bmatrix}, \tag{6.2}$$

where the a_{ij} and ϕ_{ij} are real. (Note that a symmetric matrix has been assumed for the present discussion; hence, either the target is a perfect conductor or only the backscattering case is being considered.) By considering the four pairs $(\hat{p}, \hat{q})$ that may be formed from the vectors

$$\begin{bmatrix} 0 \\ 1 \end{bmatrix}, \begin{bmatrix} 1 \\ 0 \end{bmatrix}, [1, 0], [0, 1]$$

one observes that $a_{ij} = (\sigma_{ij})^{1/2}$ where σ_{ij} is the apparent cross-section for receiving and transmitting antennas in the $\hat{i}$ and $\hat{j}$ directions, respectively, and that the ϕ_{ij} are the corresponding phase changes. The σ_{ij} with $i \neq j$ are called cross-polarization cross-sections. The six real numbers in the symmetric matrix S may be reduced to five by factoring out one of the phase angles, since the phase angles are measured relative to an arbitrary reference. The cross-section indicated by equation 6.1 is not changed by this modification.

Although the scattering matrix that has been discussed so far is applicable to bistatic as well as monostatic radar systems, the monostatic case is normally of principal interest. Therefore, we will restrict the following discussion to the monostatic case to simplify the geometry. If bistatic operation should later be adopted, the treatment will have to be generalized.

For the backscattering case we adopt the coordinate system shown in the sketch. Without loss of generality, $\hat{p}$, $\hat{q}$, and S may be written:

$$\hat{p} = \sin\psi_t \hat{i} + \cos\psi_t e^{i\delta_t}\hat{j}, \tag{6.3}$$

$$\hat{q} = \sin\psi_r \hat{i} + \cos\psi_r e^{i\delta_r}\hat{j}, \tag{6.4}$$

$$S = \begin{bmatrix} a_{11} & , & a_{12}e^{i\theta} \\ a_{12}e^{i\theta} & , & a_{22}e^{i\phi} \end{bmatrix}. \tag{6.5}$$

For time dependence e^{iwt} and a wave of polarization $\hat{p} = \sin\psi\hat{i} + \cos\psi e^{i\delta}\hat{j}$ propagating in the direction indicated in figure, the following polarization terminology obtains:

$0 < \delta < \pi$ left elliptical polarization,

$\pi < \delta < 2\pi$ right elliptical polarization,

$\delta = \frac{\pi}{2}, \omega = \frac{\pi}{4}$ left circular polarization,

$\delta = -\frac{\pi}{2}, \psi = \frac{\pi}{4}$ right circular polarization.

These definitions are in accordance with the nonclassical definitions established by IRE (1942) *Standards on Radio Wave Propagation, Definitions and Terms*—that is, that right polarization is a clockwise wave receding or a counterclockwise wave approaching. From equations 6.1, 6.3, 6.4, and 6.5, we have

$$\sigma_{pq} = | a_{11} \sin \psi_t \sin \psi_r + A_{12} \cos \psi_t \sin \psi_r e^{i(\theta+\delta_t)}, \\ + a_{12} \sin \psi_t \cos \psi_r e^{i(\theta+\delta_r)} + a_{22} \cos \psi_t \cos \psi_r e^{i(\phi+\delta_t+\delta_r)} |^2. \quad (6.6)$$

This equation is valid for backscattering from an arbitrary body consistent with the assumptions in equation 6.1. Equation 6.6 may be expanded to become (using the relationship $(a_{ij})^2 = \sigma_{ij}$):

$$\begin{aligned}\sigma_{tr} = {} & \sigma_{11} \sin^2 \psi_t \sin^2 \psi_r \\ & + \sigma_{12}[\cos^2 \psi_t \sin^2 \psi_r + \sin^2 \psi_t \cos^2 \psi_r \\ & + \tfrac{1}{2} \sin 2\psi_t \sin 2\psi_r \cos(\delta_t - \delta_r)] + \sigma_{22} \cos^2 \psi_t \cos^2 \psi_r \\ & + \tfrac{1}{2}\sqrt{\sigma_{11}\sigma_{22}} \sin 2\psi_t \sin 2\psi_r \cos(\phi + \delta_r + \delta_t) \\ & + \sqrt{\sigma_{11}\sigma_{22}}[\sin 2\psi_t \sin^2\psi_r \cos(\theta + \delta_t) + \sin^2\psi_t \sin 2\psi_r \cos(\theta + \delta_r)] \\ & + \sqrt{\sigma_{12}\sigma_{22}}[\cos^2 \psi_t \sin 2\psi_r \cos(\theta - \phi - \delta_r) \\ & + \sin 2\psi_t \cos^2 \psi_r \cos(\theta - \phi - \delta_t)]. \end{aligned} \quad (6.7)$$

It should be noted that equation 6.7 is unchanged if transmitting and receiving antennas are interchanged—that is, if ψ_r, δ_r and ψ_t, δ_t are interchanged. This result is a consequence of the reciprocity theorem or of the symmetry of the S matrix.

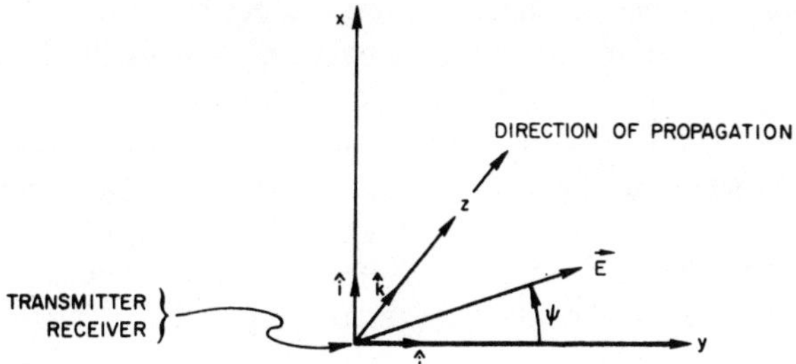

COORDINATE SYSTEM

It is of interest to determine the elements of the scattering matrix or, equivalently, of σ_{11}, σ_{12}, σ_{22}, θ, ϕ by means of radar cross-section measurements. One method of obtaining the scattering matrix originally suggested by the University of Michigan [32] has shown merit. Five cross-section measurements are made, with the radar transmitting and receiving antennas having the following linear polarizations.

	Transmitter	*Receiver*
1.	Horizontal	Horizontal
2.	Vertical	Vertical
3.	Vertical	Horizontal
4.	Linear, at any arbitrary angle (A) to horizontal	Horizontal
5.	Linear, at any arbitrary angle (A) to horizontal	Vertical

The columns headed *Transmitter* and *Receiver* may, of course, be interchanged in accordance with the reciprocity principle.[4]

Although it is stated in [32] that the complete scattering matrix may be so determined, such is not the case. The five measurements give σ_{22}, σ_{11}, σ_{12}, $\cos(\theta)$, $\cos(\theta - \phi)$. Because $\cos(-\theta) \equiv \cos(\theta)$, the values of θ and ϕ that may be found by means of linearly polarized measurements cannot be determined unambiguously. If only linearly polarized cases are to be considered, the results of the five measurements are sufficient for any case that may arise. If elliptical- or circular-polarization results are desired, two additional cross-section measurements utilizing a circularly polarized antenna are required to resolve the ambiguous signs. The fact that the method stated in [32] is incomplete and does not in fact give the complete scattering matrix, was determined independently at this Laboratory and at the University of Michigan.[5]

If only linearly polarized antennas are available, the resolution of the ambiguous signs of (θ) and $(\theta - \phi)$ is not necessary, and the five cross-section measurements are sufficient. It must be emphasized that the set of five measurements must be made at every aspect angle at which the target may be viewed by a radar; therefore, a great deal of data will be required. For a general body several feet long viewed at C-band, some 50,000 aspect angles may be necessary; the exact number of measurements required depends on the shape of the body as well as on its size and the accuracies desired.

It is possible that a large target might tend to return a radar wave with the same polarization as the incident wave. In that case the scattering matrix would not be required, and a single cross-section measurement at each aspect angle would suffice.

Great simplification results when the target is a body of revolution. It is then possible to rotate the coordinate system so that the axis of revolution of the target is in the $\hat{j}$-$\hat{k}$ plane of the figure. As a result, the scattering matrix is required for aspect angles from nose-on to tail-on in only one plane, rather than in the many planes required for the nonsymmetrical target. Measurements can therefore be made during a single rotation of the target for each of the five measurements required.

If the target is a perfect conductor in addition to being rotationally symmetric, it can be shown that the scattering matrix becomes diagonal ($\sigma_{12} = 0$) in the coordinate system named earlier. (Strictly speaking, the target need not even be rotationally symmetric; all that is required is that the $\hat{j}$-$\hat{k}$ plane be a plane of symmetry of the target and that the target be a perfect conductor.) Equation 6.7 then reduces to:

$$\sigma_{tr} = \sigma_{11} \sin^2 \psi_t \sin^2 \psi_r + \sigma_{22} \cos^2 \psi_t \cos^2 \psi_r + \tfrac{1}{2}\sqrt{\sigma_{11}\sigma_{22}} \sin 2\psi_t \sin 2\psi_r \cos(\phi + \delta_t + \delta_r). \tag{6.8}$$

The elements of the scattering matrix can now be found from four measurements: three linearly polarized measurements to determine σ_{11}, σ_{22}, and $\cos \phi$, and a circularly polarized measurement to determine the sign of ϕ.

Although many targets appear to be rotationally symmetric, not all objects of interest have this characteristic. Slot antennas for telemetry purposes and various attachments to the rear, such as hatches, render the targets asymmetric. Furthermore, the surface materials are not necessarily perfectly conducting. For these reasons it appears likely that measurements in a single plane, though sufficient for a true body of revolution, may be insufficient for a nose cone. It is therefore also probable that the simplified expression of equation 6.8 is not valid and that the five-measurement system (seven-measurement system if nonlinear polarization is to be considered) and equation 6.7 are required. Because of the analytical complexities involved, it is not possible to determine whether or not the more complex treatment is always necessary; only by a series of measurements can the situation be evaluated. For example, if measurements made on the nose cone using the five-measurement system indicated that the value of σ_{12} was much less than σ_{11} and σ_{22} over a particular range of angles, then the simplified treatment (equation 6.8) would suffice over that range of angles.

Determination of the scattering-matrix elements from measurements of radar cross-section alone is rendered difficult by errors that occur in the measurements. The magnitudes of the values of $\cos(\theta - \phi)$ and of $\cos(\theta)$ may be greater than unity as a result of measurement errors. It is then not possible to find $(\theta - \phi)$ or (θ) from their cosines. If only linear polarization is being considered, this difficulty does not arise because only $\cos(\theta - \phi)$ and $\cos(\theta)$ are needed. Values of the cosines in excess of unity, then, do not render computation impossible but do, of course, lead to incorrect results.

To investigate some effects of measurement errors on the determination of scattering-matrix elements, a simplified case has been chosen. It has been assumed that the target is a symmetric, perfectly conducting body so that equation 6.8 is a complete representation of the scattering properties of the target. Linear polarization has been assumed, since circular polarization is of no immediate interest.

Measurements of radar cross-section are made (at any particular aspect angle) with linearly polarized antennas. Transmitting and receiving antennas have the same polarization; and measurements are made with horizontal polarization, vertical polarization, and 45° polarization (that is, midway between horizontal and vertical polarization). Then, from equation 6.8 we can show that:

$$\sigma_{22} = \sigma_h, \tag{6.9}$$

$$\sigma_{11} = \sigma_v, \tag{6.10}$$

$$\cos\phi = \frac{2\sigma_{45} - \frac{1}{2}(\sigma_h + \sigma_v)}{\sqrt{\sigma_h \sigma_v}}, \tag{6.11}$$

where σ_v, σ_h and σ_{45} are the cross-sections measured with antenna polarized vertically, horizontally and at 45°, respectively. From the requirement that $-1 \leq (\cos\phi) \leq 1$, we have:

$$\tfrac{1}{4}(\sqrt{\sigma_h} - \sqrt{\sigma_v})^2 \leq \sigma_{45} \leq \tfrac{1}{4}(\sqrt{\sigma_h} + \sqrt{\sigma_v})^2. \tag{6.12}$$

The value of σ_{45} cannot be restricted any further unless definite restrictions are placed on the target shape.

From equation 6.8, 6.11, and 6.12 several important conclusions can be reached. If $\sigma_v = \sigma_h$, then $\sigma \leq \sigma_{45} \leq \sigma_h$. The cross-polarization cross-section σ_{cp} (that measured if $\psi_r = \psi_t \pm \pi/2$) is given by:

$$\sigma_{cp} = [\tfrac{1}{2}(\sigma_h + \sigma_v) - \sigma_{45}]\sin^2 2\psi_t. \tag{6.13}$$

Thus, if ψ_t (or ψ_r) is zero or 90°, the cross-polarization cross-section is zero regardless of the target properties.[6] Also, if $\sigma_{45} = (1/2)(\sigma_h + \sigma_v)$, the cross-polarization cross-section is zero for any ψ. For σ_{cp} to be greater than zero, σ_{45} must be less than $(1/2)(\sigma_h + \sigma_v)$; this requirement is fulfilled if σ_{45} obeys the inequality of equation 6.12 because $(1/2)(\sigma_h + \sigma_v) \geq (1/4)(\sqrt{\sigma_h} + \sqrt{\sigma_v})^2$ holds.

Suppose the elements of the scattering matrix necessary to permit use of equation 6.8 were to be determined by means of cross-section measurements and equation 6.9, 6.10, and 6.11, but that the measured values of cross-section were incorrect so that instead of σ_{45}, a value $\sigma_{45}(1 + x_1)$ was measured. Similarly, suppose σ_h was measured as $\sigma_h(1 + x_2)$ and σ_v as$\sigma_v(1 + x_3)$. Let us also assume for simplicity that $\psi_t = \psi_r = \psi$. Then for linear polarization equation 6.8 becomes:

$$\sigma_t = \sigma_v \sin^4\psi + \sigma_h \cos^4\psi + \sin^2\psi \cos^2\psi[4\sigma_{45} - (\sigma_h + \sigma_v)], \quad (6.14)$$

where σ_t is the true cross-section of the target. Combining terms gives

$$\sigma_t = \sigma_v \sin^2\psi(\sin^2\psi - cos^2\psi) - \sigma_h \cos^2\psi(\cos^2\psi - \sin^2\psi) \\ + 4\sigma_{45} \sin^2\psi \cos^2\psi. \quad (6.15)$$

If we let

$$A \triangleq \sin^2\psi(\sin^2\psi - \cos^2\psi), \quad (6.16)$$

$$B \triangleq \cos^2\psi(\cos^2\psi - \sin^2\psi), \quad (6.17)$$

$$C \triangleq 4 \sin^2\psi \cos^2\psi, \quad (6.18)$$

the fractional error in the cross-section for arbitrary polarization ψ, σ_m, as determined from the erroneous measurements of σ_h, σ_v, σ_{45}, is then

$$\varepsilon \triangleq \frac{\sigma_m - \sigma_t}{\sigma_t} = \frac{x_1\sigma_v A + x_2\sigma_h B + x_5\sigma_{45} C}{\sigma_v A + \sigma_h B + \sigma_{45} C}. \quad (6.19)$$

If it is assumed that x_i is a normally distributed independent random variable of mean m_i and variance V_i^2, the mean-square error $\overline{\varepsilon^2}$ is:

$$\overline{\varepsilon^2} = \frac{\left\{\begin{matrix}(V_1^2 + m_1^2)\sigma_v^2 A^2 + (V_2^2 + m_2^2)\sigma_h^2 B^2 + (V_3^2 + m_3^2)\sigma_{45}^2 C \\ + 2m_1m_2\sigma_v\sigma_h AB + 2m_1m_3\sigma_v\sigma_{45} AC + 2m_2m_3\sigma_h\sigma_{45} BC\end{matrix}\right\}}{(\sigma_v A + \sigma_h B + \sigma_{45} C)^2}. \quad (6.20)$$

As it stands, this expression is too complicated for convenient interpretation; so its application to certain special cases has been examined to learn how serious the measurement errors may be in their effects on other cross-sections computed from the set of measured values. It is reasonable to assume that $V_1^2 = V_2^2 = V_3^2 = V^2$ holds, because the randomness of measurements on a particular system should not change from one measurement to the next. Also, let $m_1 = m_2 = m_3 = 0$. Then:

$$\bar{\varepsilon} = V \frac{\sqrt{\sigma_v^2 A^2 + \sigma_h^2 B^2 + \sigma_{45}^2 C^2}}{|\sigma_v A + \sigma_h B + \sigma_{45} C|}. \qquad (6.21)$$

Because A or B may be negative, the value of $\bar{\varepsilon}$ may be less than, equal to, or greater than the value of V. Rewriting equation 6.21 gives

$$\bar{\varepsilon} = V \frac{\sqrt{A^2 + \left(\dfrac{\sigma_h}{\sigma_v}\right)^2 B^2 + \left(\dfrac{\sigma_{45}}{\sigma_v}\right)^2 C^2}}{\left| A + \dfrac{\sigma_h}{\sigma_v} B + \dfrac{\sigma_{45}}{\sigma_v} C \right|}. \qquad (6.22)$$

The expression is symmetrical about $(\sigma_h/\sigma_v = 1)$ since the A and B functions are similar in shape, and equation 6.21 could equally well have been divided through by σ_h. If $\sigma_h/\sigma_v > 10$, $\bar{\varepsilon} \approx V$; similarly, if $\sigma_v/\sigma_h > 10$, $\varepsilon \approx V$. If $\sigma_h/\sigma_v < 10$, the expression becomes more complicated to evaluate; but it appears that the rms error will seldom exceed V by very much. If the mean value of the measurement error is not zero, the effects of measurement errors may be much more serious. A detailed set of graphs similar to those used in chapter 4 to evaluate certain cross-section measurement systems would be necessary to establish the measurement accuracy required to obtain the scattering-matrix elements with any desired accuracy. Such plots may be prepared using equation 6.20. It is anticipated that such plots, when made, will indicate the need for very precise measurements.

An error analysis similar to the foregoing could well be carried out for the more general case of equation 6.7 as well; although such an analysis has not been made, the methods discussed for equation 6.8 would apply equally well to equation 6.7.

The general scattering matrix and some of its simpler, restricted forms have been discussed in the preceding pages. The possibility of determining the scattering matrix by means of radar cross-section measurements alone has been discussed in some detail. An alternative method of studying the polarization properties of radar targets involves the use of a trimode turnstile junction with appropriate electronic instrumentation to determine the relative phases of orthogonal components of the scattered wave.[7] The relative merits of such a system have been investigated in considerable detail, and it is believed that definite advantages might be gained from such direct measurements of RF phase angles.

Notes

1. Since the polarization vectors may be complex, $\hat{p}$ as a unit vector means that $\hat{p} \cdot \hat{p}^* = 1$, where * indicates complex conjugate.
2. In equation 6.1 it is assumed that polarization of a receiving antenna means the polarization of that antenna when it is used as a transmitting antenna.
3. Strictly speaking, tensor rather than matrix notation should be used to describe radar scattering; but since no advantage would accrue from such notation, the scattering matrix, a more familiar concept, is used.
4. Although separate transmitting and receiving antennas are mentioned, the discussion is still directed at a monostatic radar system; the separate antennas are assumed to be very close together and to be very far from the target, so that the bistatic angle approaches zero.
5. J. Crispin, University of Michigan, personal communication.
6. Except for the previously noted restrictions on body shape. This rather surprising result has been verified experimentally.
7. P.J. Allen and R.D. Tompkins, "An Instantaneous Microwave Polarimeter," *Proc. IRE* (July 1959): 1231–1237.

References

1. Glazier, E.V.D.; Rechtin, E.; and Voge, J., eds. *Avionics Research: Satellites and Problems of Long Range Detection and Tracking*. London: Pergamon Press, 1960, pp. 188–235.
2. *Project SAMMI-Radar Signature Analysis Program*. Contract AF 19(628)-536, Interim Technical Report Phase I, Radio Corporation of America, Moorestown, N.J., 15 March 1963.
3. *Summary Report on the Evaluation of Scintillation and Polarization Radar Signatures*. CAL Report UB-1376-S-103, Cornell Aeronautical Laboratory, Buffalo, October 1961 (Contract BTL P.O. D-292919) (S).
4. Thompkins, R.D. *An Approach to Radar Target Classification Through Polrization-Simulation Techniques*. Report 5579, Naval Research Laboratory, Washington, D.C., 23 January 1961.
5. Huynen, J.R. "Study on Ballistic Missile Sorting Based on Radar Cross-Section Data." Report no. 4, *Radar Target Sorting Based upon*

Polarization Signature Analysis. LMSD-288216, Lockheed Aircraft Corporation, Missiles and Space Division, Sunnyvale, Calif., May 1960.

6. Mikulski, J.J. *The Scattering Matrix: Polarization Power and Periodic Bodies from the Viewpoint of Matrix Theory.* Report 47.38, Massachusetts Institute of Technology, Lincoln Laboratory, Cambridge, Mass., 1 March 1960.
7. Millman, G.H. *Atmospheric and Extraterrestrial Effects on Radio Wave Propagation.* Technical Information Series no. R60EMH36, General Electric Company, Heavy Military Electronics Department, Syracuse, N.Y., June 1960.
8. Kennaugh, E.M. *Polarization Properties of Radar Reflections.* Report 389-12 (AD2494), Ohio State University Research Foundation, Antenna Laboratory, Columbus, 1 March 1952.
9. Copeland, J.R. "Range Target Classification by Polarization Properties." *Proc. IRE* 48(1960): 1290–1296.
10. Crispen, J.W., et al. *The Measurement and Use of Scattering Matrices.* Report no. 2500-3-T, University of Michigan, Radiation Laboratory, February 1961.
11. Bechtel, M.E.; Melling, W.P.; and Graham, G.A.R. *Final Report, Project ACRE.* Report UB-1405-P-1, Cornell Aeronautical Laboratory, Buffalo, May 1960 (Contract RCA P.O. GX9-947350-0001-96-Q78).
12. Longacre, A., and Waller, A. *Non-coherent Doppler Detection of Vibratory Motion.* Final Report DSL-R-72, Syracuse University Research Corporation, 15 December 1961 (Contract NOBSR).
13. Booker, H.G., et al. "Techniques for Handling Elliptically Polarized Waves with Special Reference to Antennas." *Proc. IRE* 39(1951): 533–556.
14. Bachman, C.G., et al. *A Study of Nanosecond Pulse Techniques in Radar Transmission.* RADC-TR-61-247 (ASTIA 268880 and 327147), CAL Report UB-1426-P-1, Cornell Aeronautical Laboratory, Buffalo, N.Y., 27 February 1961.
15. Klauder, J.R., et al. "The Theory and Design of Chirp Radars." *Bell System Tech. J.* 39(1960).
16. Key, E.L., et al. *A Method of Pulse Compression Employing Nonlinear Frequency Modulation.* MIT-LL TR-207, Massachusetts Institute of Technology, Lincoln Laboratory, Cambridge, Mass., 13 August 1959.
17. Woodward, P.M. *Probability and Information Theory, with Applications to Radar.* London: Pergamon Press, 1953.
18. Dunn, V.E. *A Pulse Compression Filter Employing a Microwave*

Helix. Technical Report no. 557-1, Stanford University, Electronics Laboratory, 27 October 1960.

19. Dunn, V.E. *Realization of Microwave Pulse Compression Filters by Means of Folded Tape Meander Lines,* Technical Report no. 557-3 (RADS-TDR-62-551), Stanford University, Electronics Laboratory, October 1962.
20. Allen, J.L. *A Quantitative Examination of the Radar Resolution Problem.* MIT-LL TR-281, Massachusetts Institute of Technology, Lincoln Laboratory, Cambridge, Mass., 17 September 1962.
21. *Superresolution—An Advanced Form of Pulse Compression.* Hazeltine Corporation, Research Division, Little Neck, N.Y., October 1963.
22. *Six Dimensional Radar Resolution: A Summary.* Hazeltine Corporation, Research Division, Little Neck, N.Y., September 1963.
23. Holmes, D.D.; Cope, R.H.; and Strauss, H. *Technical Proposal for Space Object Identification Study.* Report no. 7742 (RADC PR 63-462). Hazeltine Corporation, Research Division, Little Neck, N.Y., April 1963.
24. *Application of Superresolution Techniques in the Trinidad Radar.* Report 7671 (Excerpt for RADC). Hazeltine Corporation, Research Division, Little Neck, N.Y., November 1960.
25. *AMR Station Trinidad Measurement Capability.* EHM-1440 (100) 337-013. General Electric Company, Heavy Military Electronics Department, Syracuse, N.Y., October 1963.
26. Bias, G.H., Jr., and Taylor, J.W., Jr. *An FPQ-4 Radar Modified for Synthetic Spectrum.* Report no. 86-20080-63, Westinghouse Electric Corporation, Electronics Division, Baltimore, Md.
27. Shoenberger, G.C. *High Resolution Range Profiles of Distributed Targets as Observed with AN/FPS 55 Synthetic Spectrum Radar.* Westinghouse Electric Corporation, Electronics Division, Baltimore, Md.
28. Goldberger, M.L.; Lewis, H.W.; and Watson, K.M. *HB-T Primer.* IDA Log S-2608/57, Institute for Defense Analyses, Washington, D.C.
29. *Proposal to Conduct Wide-Baseline Correlation Interferometer Experiments on Satellite Inspection.* Proposal no. 233 (for ARPA), Cornell Aeronautical Laboratory, Buffalo, N.Y., 28 June 1963.
30. Brown, R.H., and Twiss, R.Q. "A New Type of Interferometer for Use in Radio Astronomy." *Phil. Mag. (Ser. 7)* 45(1954): 366.
31. *Automatic Target Recognition Analysis Program (AUTRAN).* Interim Report ESD-TDR-63-621, Radio Corporation of America, Moorestown, N.J., October 1963 (Contract AF 19(628)-580).
32. Crispin, J.W., Jr.; Goodrich, R.F., and Siegel, K. *A Theoretical Method for the Calculation of the Radar Cross Section of Aircraft and*

Missiles, Report no. 2591-H, University of Michigan, Radiation Laboratory, July 1959 (Contract AF 19(604)-1949).

33. Evans, G.W., II, et al., *Objectives and Plans for Mathematical Investigations of Radio Frequency Backscatter from Missiles.* Stanford Research Institute, Menlo Park, Calif., March 1962 (Contract SD-103; ARPA Order 281-62; Project Code 7400).
34. Peterson, A.M. "Upper Atmosphere Clutter Research." *Effects of the Atmosphere on Radar Resolution and Accuracy.* RADC-TR-60-44, Stanford Research Institute, Menlo Park, Calif., April 1960 (Contract AF 30(602)-1762).
35. Winograd, S., and Cowan, J.D. *Reliable Computation in the Presence of Noise.* Cambridge, Mass.: MIT Press, 1963.

7 Theory of the Hanbury-Brown–Twiss (HB-T) Technique

Introduction

The use of spaced receivers to determine the size of ionospheric or celestial objects is well documented in the radiophysics literature. Recently there has been some interest in the application of these techniques to the problem of determining the size of orbiting objects by ground-based radar observations. In particular, attention has been focused on the process of amplitude-only measurements at separated receiving points similar to the Hanbury-Brown and Twiss *correlation interferometer* used in radio astronomy.

In this correlation interferometer, amplitude data are obtained at separated receiving sites. What is the expected output from such a correlation interferometer if the scattering object is a satellite illuminated by a ground-based transmitter?

Spatial-Correlation Techniques

For simplicity, consider a two-dimensional geometry with the satellite located a distance of R_0 above a baseline of length r betwen the two receiving antennas. Further assume that the position of the satellite is fixed with respect to the normal to the antenna baseline and that its tumbling motion can be represented adequately by rotating the object about its center of gravity. The geometry of this situaiton is illustrated in figure 7–1. Define $A(\alpha, \theta)$ as the scattering-amplitude polar diagram of the satellite when illuminated by a transmitter in a direction α from the reference axis; θ is the angle from the reference axis of the satellite to the receiver. Under these conditions we have

$$E_1 = A(\alpha, \theta_1)[\omega t + \phi_1(t)],$$

$$E_2 = A(\alpha, \theta_2)[\omega t + \phi_2(t)],$$

where ω is the angular frequency and ϕ_1 is the phase path.

Taken directly from *Proposal to Conduct Wide-Baseline Correlation Interferometer Experiments on Satellite Inspection*. Proposal no. 233 (for ARPA), Cornell Aeronautical Laboratory, Bufalo, N.Y., 28 June 1963.

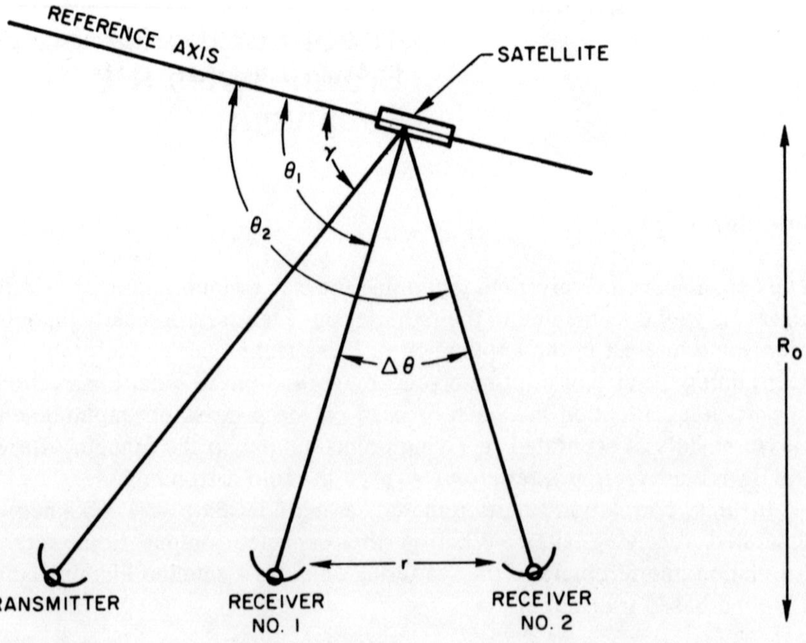

Figure 7–1. Radiation Environment Geometry for HB-T Technique

If we envelope-detect (a conventional peak detector) and cross-correlate the two receiver outputs, the normalized output of the interferometer will be

$$\frac{\overline{A\{\alpha(t),\, \theta_1(t)\}A\{\alpha(t),\, \theta_2(t)\}}}{\overline{A^2\{\alpha(t),\, \theta,\, (t)\}}}\,.$$

If the tumbling rate of the satellite is large enough so that (in an average over time) all values of α (and therefore θ_1 and θ_2) are equally probable, then the time average is equivalent to an average over all angles. Consequently,

$$E_{\text{out}} = \frac{\overline{A(\alpha(t),\, \theta_1(t))A(\alpha(t),\, \theta_2(t))}}{\overline{|\, A(\alpha(t),\, \theta_1(t)\,|^{\,2}}} = \frac{\langle A(\alpha,\, \theta_1)A(\alpha,\, \theta_1 + \Delta\theta)\rangle}{\langle\, |\, A(\alpha,\, \theta_1)|^{\,2}\rangle}\,.$$

The output is a function only of $\Delta\theta$, which from geometry is a function of r and R_0:

$$E_{\text{out}} = \rho(\Delta\theta) = \rho\left(\frac{r}{R}\right).$$

ρ is recognized as the correlation function of the amplitude-scattering diagram averaged over all angles α and θ and should be related in some way to the size of the object in wavelengths. If the object is large (in wavelengths), the amplitude-scattering diagram should exhibit narrow lobes or fine structure in the $\rho(\alpha, \theta)$. Consequently, one would expect that the correlation function would decrease rapidly for $\Delta\theta$ differing from zero. Conversely, if the object size were small (measured in wavelengths), giving rise to a broad lobe structure, one would expect the correlation function to have a significant value for small values of $\Delta\theta$.

If we used the Brown and Twiss technique of square-law detectors instead of linear detectors, it should be obvious that the output would be proportional to the correlation of the "power" polar scattering diagram. We have carried out the analysis of the output signals when the linear and/or square-law detectors are followed by high pass filters (similar to Brown and Twiss) and the signals are specified only by amplitude distributions. The outputs are shown to be proportional to the correlation between the amplitude-scattering polar diagram or its square.

In all cases of interest, the output of such a spaced receiving system is shown to be proportional to the correlation of the amplitude-scattering polar diagram. If enough receivers are spaced so that a reasonably accurate plot of $\rho(\Delta\theta)$ is obtained, then an "average" size of the object may be specified through the relationship

$$\Delta\theta_0 \sim \frac{\lambda}{D},$$

where

$\Delta\theta_0$ = the magnitude of r/R_0, which, for example, makes $\rho(\Delta\theta) = e^{-1}$,

λ = the operating wavelength,

D = the "average" size of the object.

Temporal-Correlation Studies

The previous section was concerned with the correlation of the simultaneous amplitudes at two receivers separated in space. It is fairly clear that, in addition to a correlation at zero time lag, the function $\rho(\Delta\theta, \tau)$ can be generalized to

$$\rho(\Delta\theta, \tau) = \overline{E_1(t)E_2(t+\tau)}$$

by delaying the output of one receiver relative to that of the other. If the amplitude polar scattering diagram of the object sweeps along the ground without a great deal of change in a time Δt, then it will first strike one receiver and some time later (less than Δt) strike another receiver. The output of the second receiver will closely duplicate that of the first receiver except that it will be delayed in time by an amount τ so that

$$\omega_R \tau = \Delta\theta,$$

where ω_R is the angular velocity of the diffraction pattern (tumbling rate) and $\Delta\theta$ is the angle subtended by the receiver baseline r at the satellite.

A monostatic radar system can sample the diffraction pattern of a target by observing the amplitude-time history of the radar returns, but one cannot distinguish from the monostatic time history between a small target (with wide angular lobes) rotating rapidly and a large target (with narrow angular lobes) rotating slowly. However, if the time-lagged spatial correlation function rises to a significant value at a time lag τ, one can determine the angular rotation rate and hence deduce a target size from the single monostatic time history.

It must be noted that for this technique to give reasonable results, the receiver separation and satellite tumbling motion must be so related that the diffraction pattern does not change drastically (perhaps because of high rates of tumbling along orthogonal axes) in the time required for the pattern to rotate through an angle $\Delta\theta$.

Required Analysis Determination

From the foregoing it is evident that a large number of analytical procedures could be applied to the prospective experimental data. To determine the most useful procedures, it is necessary to compute the following correlations for each receiver pair and determine how each may best be used to estimate satellite characteristics. The procedure for deriving maximum information about the satellite from correlation interferometers of similar configuration will then be delineated.

The outputs of any two receivers may be defined as $E_1(\tau)$ and $E_2(\tau)$; then we compute

$$\rho_1\left(\frac{r_{12}}{R_0}\right) = \frac{\overline{E_1(t)E_2(t)}}{\overline{|E_1(t)^2|}^{\,1/2}\{\overline{|E_2(t)|^2}\}^{1/2}}, \quad \text{(amplitude polar scattering diagram correlation coefficient)} \tag{7.1}$$

$$\rho_2\left(\frac{r_{12}}{R_0}\right) = \frac{\overline{E_1^2(t)E_2^2(t)}}{\{\overline{|E_1^2(t)|^2}\}\{\overline{|E_2^2(t)|^2}\}}, \quad \text{(power polar scattering diagram correlation coefficient)} \tag{7.2}$$

$$\rho_3\left(\frac{r_{12}}{R_0}\right) = \frac{\overline{(E_1(t) - \overline{E_1(t)})(E_2(t) - E_2(t))}}{\text{Normalizing Value}}, \quad \text{(linear-detector HB-T)} \tag{7.3}$$

$$\rho_4\left(\frac{r_{12}}{R_0}\right) = \frac{(E_1^2(t) - \overline{E_1^2})(E_2^2(t) - \overline{E_2^2})}{\text{Normalizing Value}}, \quad \text{(square-law HB-T)} \tag{7.4}$$

$$\rho_5\left(\frac{r_{12}}{R_0}, \tau\right) = \frac{\overline{E_1(t)E_2(t+\tau)}}{\{\overline{|E_1(t)|^2}\}^{1/2}\{\overline{|E_2(t)|^2}\}^{1/2}}. \quad \text{(tumbling rate time and space correlation coefficient)} \tag{7.5}$$

Limitations of the Technique

The relationship between the "average" size of the object and its real physical dimensions is not at all clear. Furthermore, the relationship between the *correlation function* of the scattering polar diagram and real physical length is not unique—many objects of widely differing shapes may give rise to the same correlation function. Even in the idealized two-dimensional geometry of a static satellite (having no translation), one can see possible limitations to the spaced-receiver or "interferometer" technique applied to satellite-size determination. When we consider three-dimensional geometry (with three-dimensional tumbling and a satellite with translation), the problem is complicated further. With three-dimensional tumbling, a time average may involve averaging a length and a width measurement. Furthermore, as the satellite progresses in its orbit, a fixed baseline separation r results in a varying included angle $\Delta\theta$ so that, in a time average of spatial variables, some averaging of $\Delta\theta$ and hence of $\rho(\Delta\theta)$ is inevitable. Other problems associated with the polarization characteristics of scattering from objects will further complicate the relationship between "size" and apparent correlation function, indicating the desirability of multistatic techniques.

8 Propagation Effects

Ionospheric Polarization Effects

Introduction

When a linearly polarized electromagnetic wave enters an ionized medium in the presence of an external magnetic field such as is encountered in the ionosphere, the wave separates into two independent components, both elliptically polarized in the general case, with opposite senses of rotation. For frequencies in the UHF range, the two components—referred to as the *ordinary* and the *extraordinary* waves—are circularly polarized.

As the ionosphere is traversed, the two waves progress with different velocities of propagation. This results in a continuously changing phase relationship between them. On leaving the ionosphere, the circularly polarized components recombine to form a linearly polarized wave that is rotated with respect to the original linear wave. This phenomenon is commonly referred to as the *Faraday effect.*

The magnitude of the angular rotation is a function of the integrated electron density along the transmission path, the angle between the direction of the propagation and the direction of the magnetic line of force, and the transmission frequency.

Theoretical Considerations

Since a circularly polarized wave is basically composed of two harmonic orthogonal signals that are 90° out of phase with each other, only one of the linearly polarized waves from each of the two circularly polarized components (that is, the ordinary and extraordinary waves) need be considered in analyzing the polarization rotation effects of the ionosphere. The electric-field intensities of two linearly polarized progressive waves can be written as

$$E_o = a \exp[j(\omega t - \gamma_1 s)], \tag{8.1}$$

Taken directly from G. H. Millman, *Atmospheric and Extraterrestrial Effects on Radio Wave Propagation*, Technical Information Series no. R60EMH36, General Electric Company, Heavy Military Electronics Department, Syracuse, N.Y., June 1960.

$$E_e = a \exp[j(\omega t - \gamma_2 s)], \tag{8.2}$$

where the subscripts o and e refer to the ordinary and the extraordinary modes of propagation, respectively. The phase constants, γ_1 and γ_2, are defined by

$$\gamma_{1,2} = \frac{\omega}{V_{p1,2}}, \tag{8.3}$$

where V_p is the phase velocity of the wave and ω is the angular transmission frequency.

Since by definition the refractive index of the media is

$$n_{1,2} = \frac{c}{V_{p1,2}}, \tag{8.4}$$

it therefore follows that

$$\gamma_{1,2} = \frac{\omega n_{1,2}}{c}, \tag{8.5}$$

where c is the velocity of electromagnetic waves in free space.

The difference in phase between the two waves traversing a distance ds is given by

$$d\phi = (\gamma_1 - \gamma_2) ds. \tag{8.6}$$

Substituting equation 8.5 in this expression, it is seen that

$$d\phi = \frac{\omega}{c} \Delta n \, ds, \tag{8.7}$$

where $\Delta n = n_1 - n_2$.

This relationship defines the phase shift for a one-way propagation path. When two-way transmission paths are considered, an additional factor of 2 is introduced. It should be noted that $d\phi$, as given by equation 8.7, is the differential phase shift between the two magnetoionic components. The resultant polarization rotation is merely one-half this value. Thus, taking these two items into consideration, the total polarization shift for a two-way propagation path becomes

$$\phi(s) = \frac{\omega}{c}\int_{s_1}^{s_2} \Delta n \, ds,$$

where s_1 and s_2 are the limits of the path.

In terms of a vertical-height variable, this expression becomes

$$\phi(h) = \frac{\omega}{c}\int_{h_1}^{h_2} \Delta n f(h) dh, \tag{8.8}$$

$$f(h) = \frac{r_o + h}{(r_0 + h)^2 - (r_0 \cos E)^{2\,1/2}}, \tag{8.9}$$

where r_0 is the radius of the earth and E is the elevation angle of the antenna beam. The parameter, $f(h)$, is plotted in figure 8–1 as a function of height. It

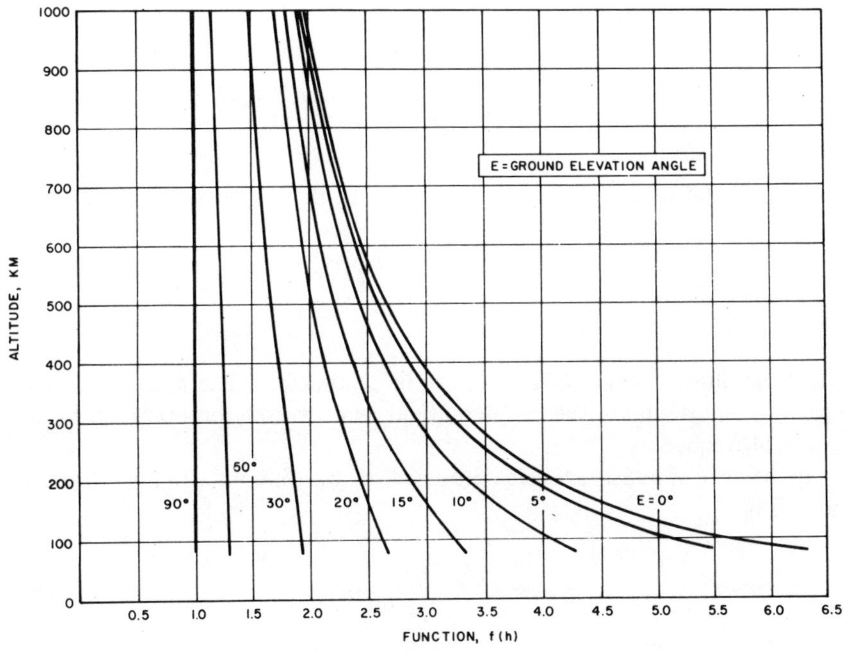

Figure 8–1. Ionospheric Function, $f(h)$

is seen that $f(h)$ tends to approach unity as the height or elevation angle is increased.

According to the theory of propagation of electromagnetic waves, the difference in the refractive indexes between the ordinary and the extraordinary waves for quasi-longitudinal propagation is given by

$$\Delta n = \frac{N_e e^3 H}{2\pi^2 m^2 c f^3} \cos\theta, \tag{8.10}$$

where N_e = the electron density (electrons/cm^3);
H = the earth's magnetic field (gauss);
e = the electron charge (4.8×10^{-10} esu);
m = the electron mass (9.1×10^{-28} gm);
f = the transmission frequency (Hz).

The total two-way polarization rotation (in radians) in substituting equation 8.10 into equation 8.8 becomes

$$\phi(h) = \frac{e^3}{\pi m^2 c^2 f^2} \int_{h_1}^{h_2} f(h) H \cos\theta N_e dh. \tag{8.11}$$

In terms of the values of the known constants, this expression reduces to

$$\phi(h) = \frac{4.7233 \times 10^4}{f^2} \int_{h_1}^{h_2} f(h) H \cos\theta N_e dh, \tag{8.12}$$

where the altitude limits h_1 and h_2 are in centimeters. The total rotation is directly proportional to the magnetic field and inversely proportional to the square of frequency.

In the case of quasitransverse propagation, the difference in the refractive indexes is

$$\Delta n = \frac{N_e e^4 H^2}{8\pi^3 m^3 c^2 f^4} \sin^2\theta, \tag{8.13}$$

where $89.8° \leq \theta \leq 90°$ for frequencies in the 100-MHz/sec range.

The two-way polarization rotation, therefore, is

$$\phi(h) = \frac{e^4}{4\pi^2 m^3 c^3 f^3} \int_{h_1}^{h_2} f(h) H^2 \sin^2\theta N_e dh, \tag{8.14}$$

or simply

$$\phi(h) = \frac{6.6087 \times 10^{10}}{f^3} \int_{h_1}^{h_2} f(h) H^2 \sin^2\theta N_e dh. \tag{8.15}$$

It is evident that the polarization is directly proportional to the magnetic field squared and inversely proportional to the cube of frequency for this mode of propagation.

The calculations of the one-way polarization rotation, plotted in figure 8–2 as a function of elevation angle, are based on the following assumptions: target located at a height of 1,000 km, transmission frequency of 100 MHz magnetic field of 0.62 gauss, and longitudinal propagation. It evident that, assuming identical propagation conditions, the polarization shift from transmission along the horizon is approximately 3.5 times greater than that at

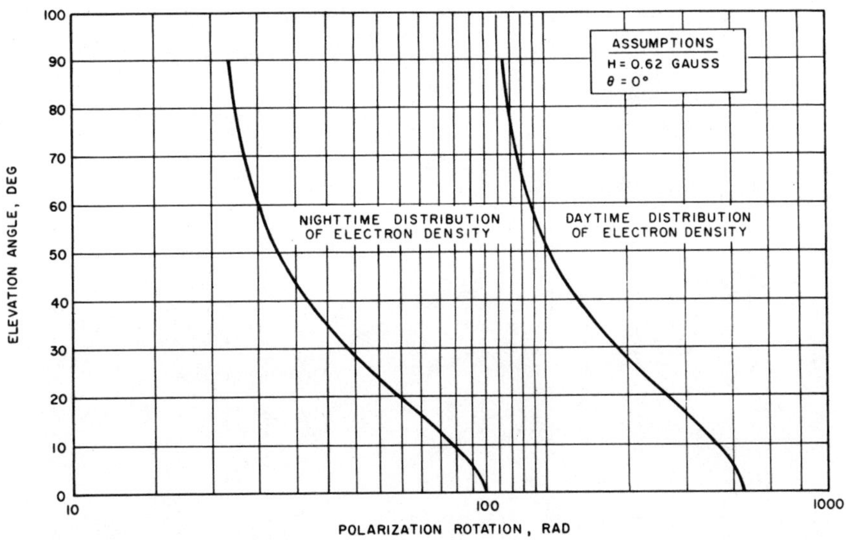

Figure 8–2. Polarization Rotation at 100 MHz for Target at 1,000-km Height, One-Way Transmission Path, and Longitudinal Propagation

an angle of 90°. In addition, for a given elevation angle, the daytime shift is on the order of 3.8 times greater than the nighttime effect.

The polarization rotation for the two modes of propagation is shown in figures 8–3 and 8–4. It is seen that for the longitudinal case the rotation is about 100 times greater than for the transverse condition.

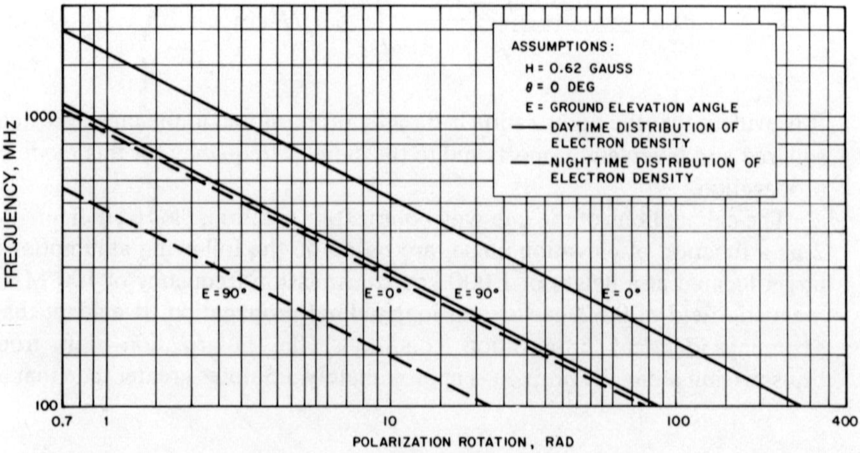

Figure 8–3. Polarization Rotation as a Function of Frequency of a Target at 1,000-km Height, One-Way Transmission Path, and Longitudinal Propagation

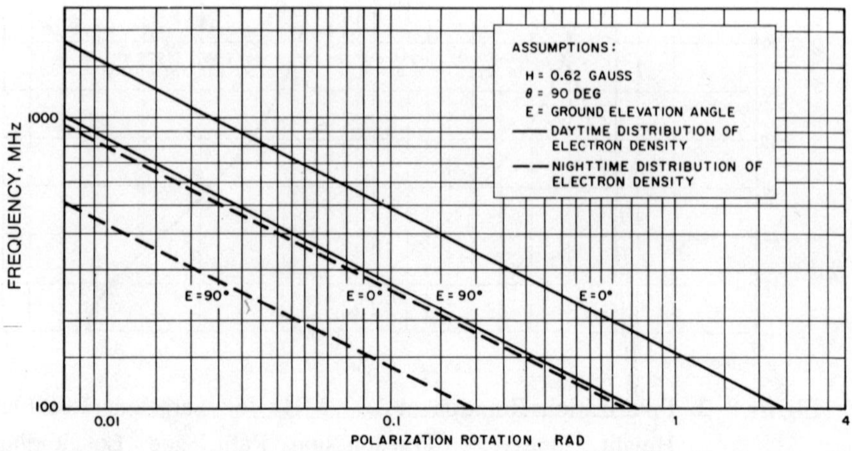

Figure 8–4. Polarization Rotation as a Function of Frequency for a Target at 1,000-km Height, One-Way Transmission Path, and Tranverse Propagation

According to figures 8–5 and 8–6, for transmissions at 0° elevation angle and daytime conditions, 95 percent of the total polarization shift is attained at a height of 550 km, but it is reduced to the 470-km level for the nighttime distribution.

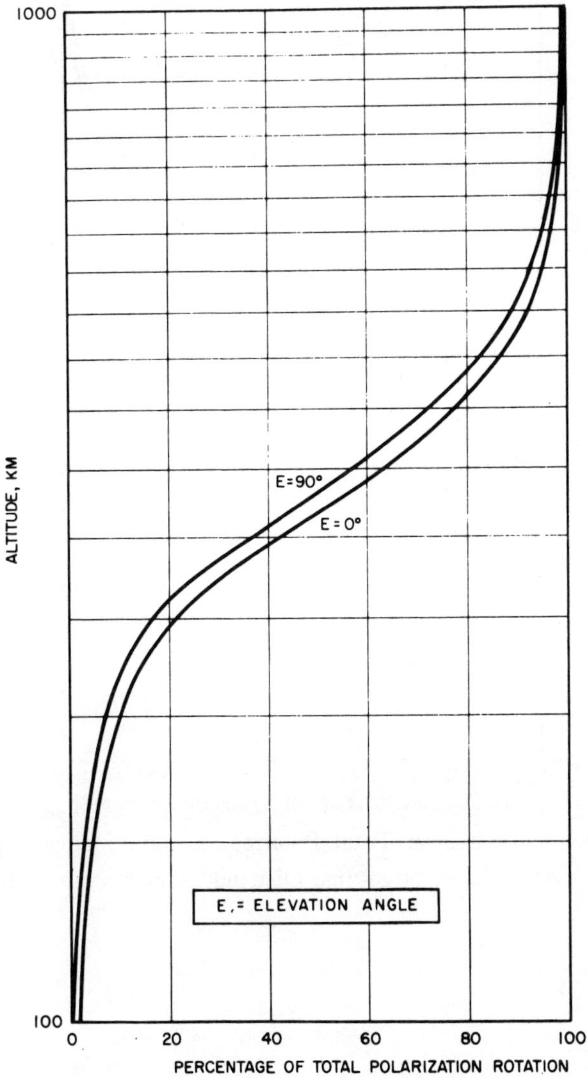

Figure 8–5. Percentage of Total Polarization Rotation as a Function of Altitude for Daytime Distribution of Electron Density

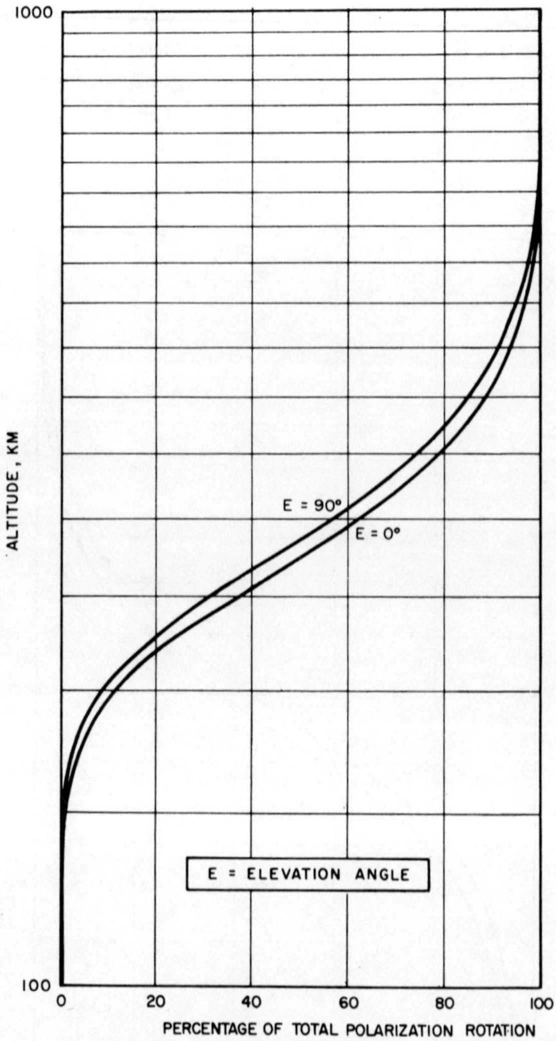

Figure 8–6. Percentage of Total Polarization Rotation as a Function of Altitude for Nighttime Distribution of Electron Density

Time Delays

Introduction

Time delays or range errors are always inherent in radar target measurements. This effect occurs primarily because the velocity of electromagnetic

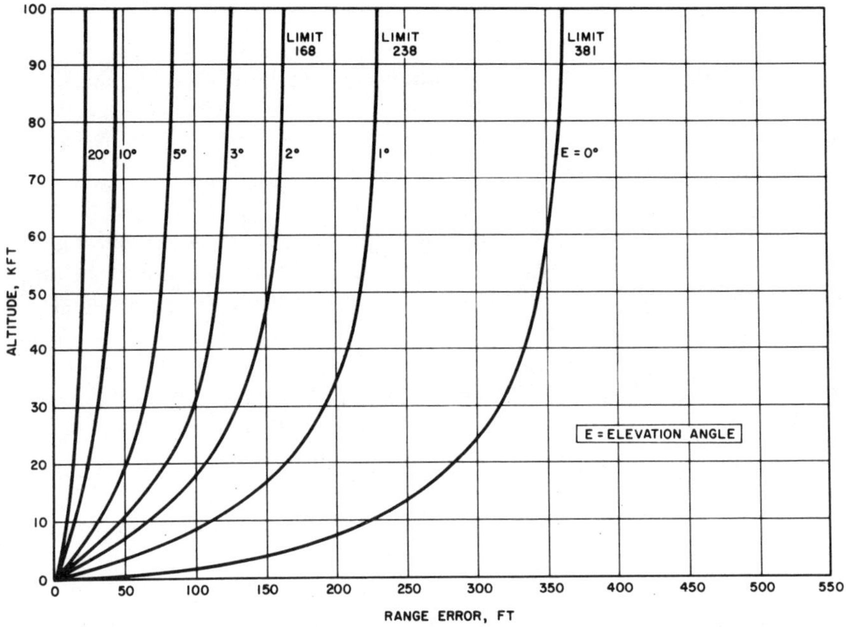

Figure 8–7. Atmospheric Range Errors for a Standard Atmosphere with 100 Percent Relative Humidity, One-Way Transmission Path

propagation in the atmosphere is slightly less than the free-space velocity. An additional source of error is the increase in path length brought about by the refractive bending of the propagated ray. In the computations (by the stratified-layer method) of the data presented in this section, both factors are taken into account.

Tropospheric Time Delay

The tropospheric range errors plotted in figures 8–7 and 8–8 are valid for two standard atmospheric models. The same number of stratified layers utilized in the tropospheric refraction calculations was also employed in this analysis.

It is seen that the magnitude of the range error is a direct function of the elevation angle or path length and also a function of the amount of humidity in the air (the value at 0 percent relative humidity is about 80 percent of the value at 100 percent relative humidity).

For a one-way transmission path, a maximum range error of approximately 381 ft is possible for propagation along the horizon. Regardless of the atmospheric model, approximately 57 percent of the retardation occurs below the 10,000-ft altitudes for horizon transmissions, whereas at 5° elevation angle and at the same altitude only 34 percent takes place.

Ionospheric Time Delay

Figures 8–9 and 8–10 are graphs of range error for daytime and nighttime models of electron density. As in the case of the ionospheric refraction computations, the base of the ionosphere—that is, the 80-km altitude for the daytime distribution and 100 km for the nighttime distribution—was taken as the point of origin of the ray trajectory.

It is evident that (1) the range error is a maximum during the daytime, increasing with low elevation angles; (2) the error is inversely proportional to the square of frequency; and (3) the major portion of the time delay occurs below 300-n mi altitude.

At 200 MHz the maximum error is on the order of 3,000 ft for a one-way transmission path. This value is approximately eight times larger than the maximum error introduced by the troposphere.

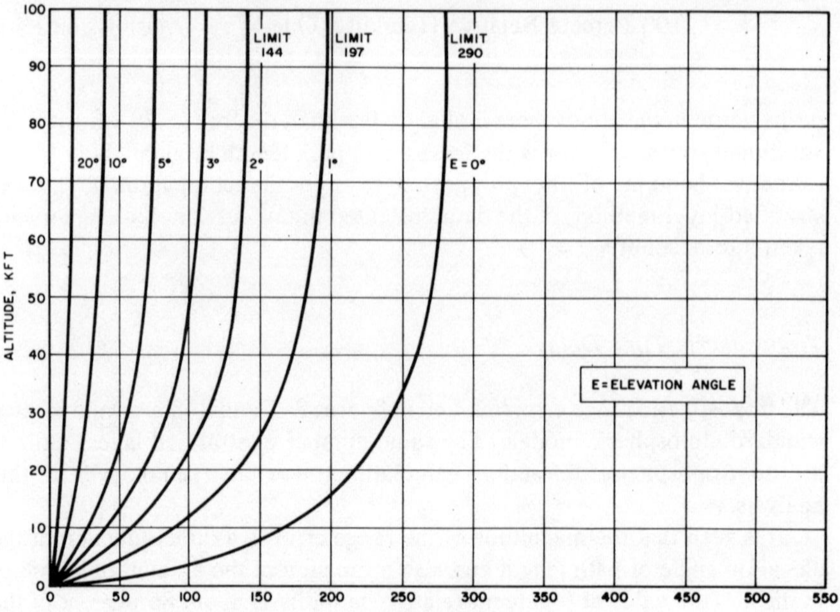

Figure 8–8. Atmospheric Range Errors for a Standard Atmosphere with 0 Percent Relative Humidity, One-Way Transmission Path

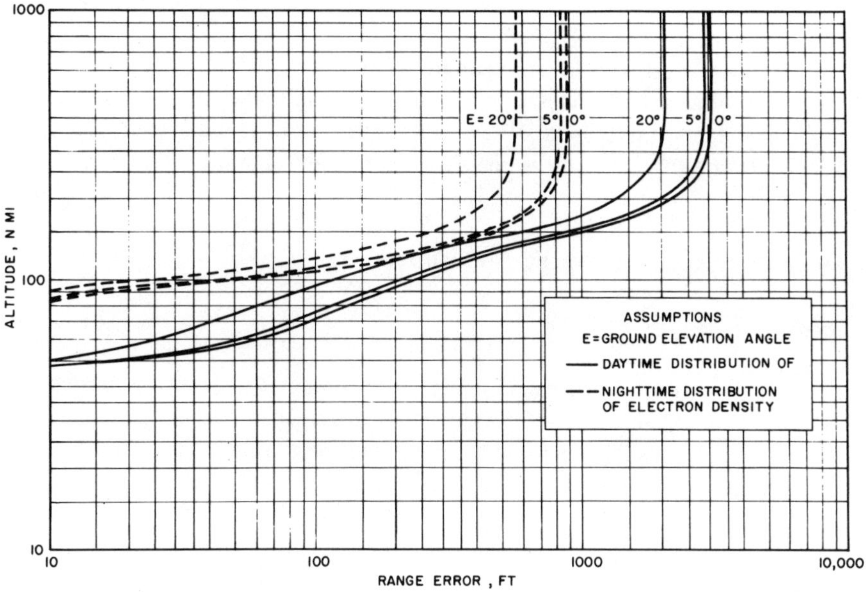

Figure 8–9. Ionospheric Range Errors at 200 MHz, One-Way Transmission Path

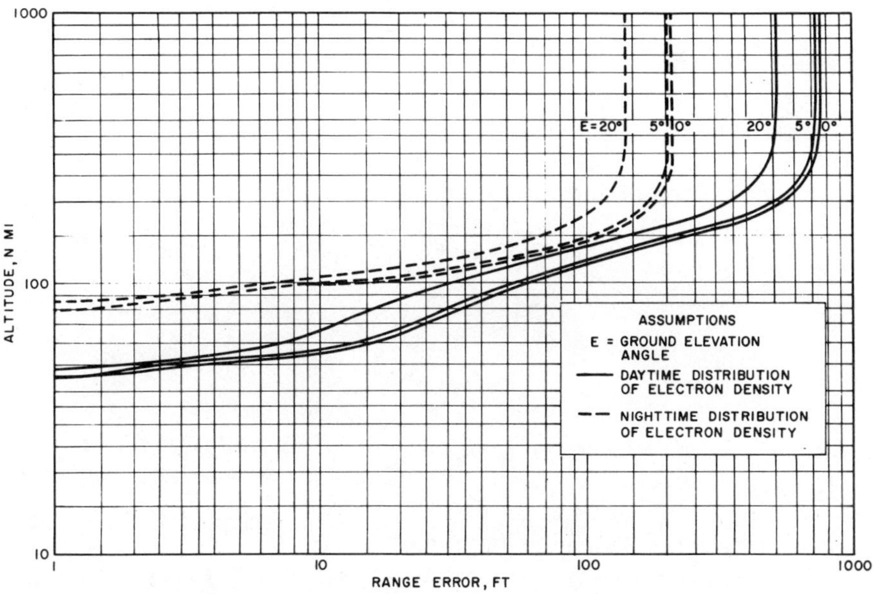

Figure 8–10. Ionospheric Range Errors at 400 MHz, One-Way Transmission Path

It follows from the inverse square of frequency dependency that, at a frequency of about 575 MHz the tropospheric and ionospheric contributions to the propagational time delays are of the same order of magnitude. At about 575 MHz the tropospheric error is the more predominant.

Doppler Effects Due to the Atmosphere

Introduction

Because of the propagational anomaly of refraction of radio waves traversing the atmosphere, an error is introduced in the determination of the line-of-sight or radial velocity of a moving target. This error results from the fact that the direction of the refracted ray at the target differs slightly from the fictitious line of sight or apparent path direction. The doppler error, introduced by the incorrect assumption of the magnitude of the refractive index at the target location in the deviating medium, is believed to be indeterminate and is therefore not considered in this analysis.

It should be mentioned that, unlike refraction and range errors, the doppler errors are not cumulative. In other words, the total doppler error for a target traversing the ionosphere is only the localized error at the specific point in the medium. The troposphere merely acts as a refractive medium

Tropospheric Effect

A target T is traveling with a velocity V in an arbitrary direction in the refractive medium. The components of target velocity for the various orientations shown in figure 8–11 are

$$V_r = V \cos(\psi + \Delta\alpha_T), \tag{8.16}$$

$$V_d = V \cos\psi, \tag{8.17}$$

$$V_a = V_d \cos(\Delta\alpha), \tag{8.18}$$

where V_r, V_d, and V_a are the components of V in the ray-path direction, direct-path direction, and apparent-path direction, respectively. The angle $\Delta\alpha$ is the retractive-error angle, whereas $\Delta\alpha_T$ is the angle between the ray path and the direct path at the target location.

The error introduced in the measurement of the target doppler velocity ΔV is given by

$$\Delta V = V_a - V_r \tag{8.19}$$

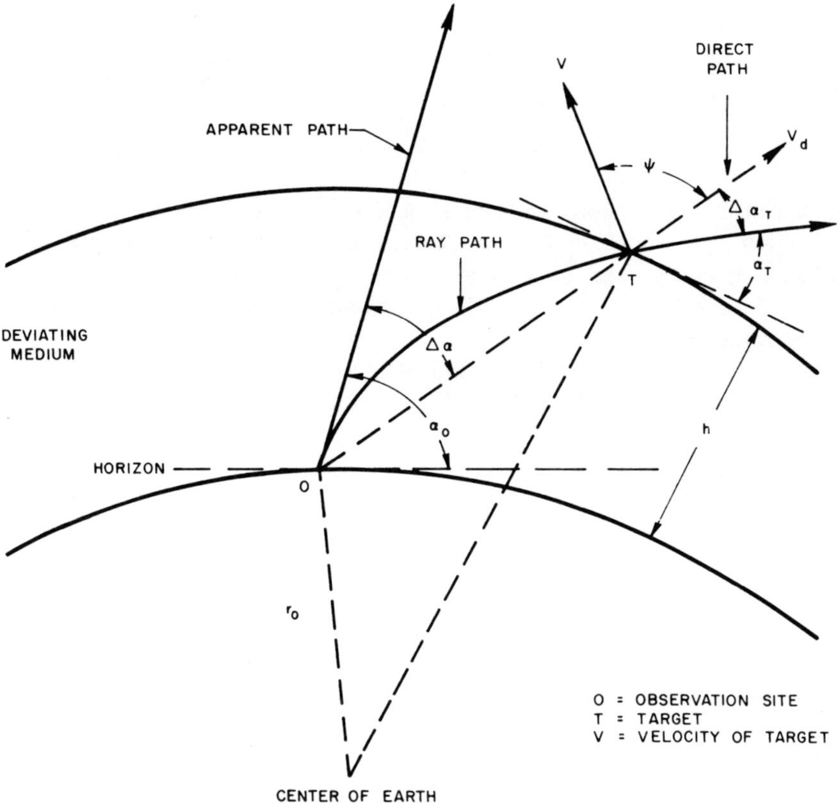

Figure 8–11. Deviation of Ray Path at Target Position

Expanding the cosine term of equation 8.16 and then substituting the resulting expression together with equations 8.17 and 8.18 into equation 8.19, it follows that the velocity error can be written in the form

$$\Delta V = V \cos \psi[\cos(\Delta\alpha) - \cos(\Delta\alpha_T)] + V \sin \psi \sin(\Delta\alpha_T) \qquad (8.20)$$

Since $\Delta\alpha$ and $\Delta\alpha_T$ are very small angles, the trigonometric functions can be expanded in a series. Neglecting higher-order terms gives

$$\Delta V \cong V \cos \psi \left[\frac{(\Delta\alpha_T)^2}{2} - \frac{(\Delta\alpha)^2}{2}\right] + V \sin \psi(\Delta\alpha_T). \qquad (8.21)$$

Since the magnitudes of $(\Delta\alpha_T)^2$ and $(\Delta\alpha)^2$ are on the order of microradians, the target radial-velocity component $V \cos \psi$ can be neglected. Thus, to a first approximation,

$$\Delta V \cong V \sin \psi(\Delta\alpha_T). \tag{8.22}$$

It is seen that the doppler velocity error in the radial direction attributed to refraction is composed only of the tangential velocity component of target velocity $V \sin \psi$. The error is a maximum when the velocity vector is perpendicular to the direction of the direct path and is a minimum when the target is moving along the direct line-of-sight path.

It is of interest to note that the identical expression for the doppler velocity error given by Equation 8.22 would have been obtained if ΔV had been defined by

$$\Delta V = V_d - V_a. \tag{8.23}$$

In other words, the magnitude of the refraction-error angle is not significant and can be neglected in this analysis.

According to the general concept of the doppler effect, the reflection of a radio wave from a moving object results in a signal returned to a fixed transmitting-receiving system whose frequency differs from the transmitted frequency. Analytically, this phenomenon is written as

$$f' = \frac{c - V \cos \psi}{c + V \cos \psi} f, \tag{8.24}$$

where f = the transmitted frequency,

f' = the apparent reflected frequency,

V = the velocity of the moving object,

c = the velocity of light in free space,

ψ = the orientation angle (see figure 8–1).

When the target is moving away from the observing site ($\psi = 0°$), the reflected frequency is less than the transmitted frequency. For an object moving toward the observer ($\psi = 180°$) the reflected frequency is greater than the transmitted frequency.

Equation 8.24 and be expressed as

$$f_d = -\frac{2V \cos \psi}{c + V \cos \psi} f, \tag{8.25}$$

where f_d is the difference between the apparent reflected frequency and the transmitted frequency. Since $c \gg V$, the frequency difference reduces to

$$f_d = -\frac{2V\cos\psi}{c} f. \qquad (8.26)$$

The error encountered in the measurement of the doppler frequency is merely

$$d(f_d) = -\frac{2f}{c}\cos\psi\, dV. \qquad (8.27)$$

This expression, of course, assumes that f and ψ are errorless quantities. The term $\cos\psi\ dV$ is the component of target-velocity error in the radial direction.

Substituting equation 8.22 in this relationship gives the doppler frequency error due to refraction as

$$d(f_d) = -\frac{2f}{c} V \sin\psi(\Delta\alpha_T). \qquad (8.28)$$

With regard to the angle $\Delta\alpha_T$, this parameter can be evaluated from Snell's law for spherically stratified surfaces and from the law of sines. It can be readily shown that

$$\Delta\alpha_T = \cos^{-1}\left[\frac{r_o}{r_o + h}\cos(\alpha_0 - \Delta\alpha)\right] - \cos^{-1}\left[\frac{n_G r_o}{n_T(r_o + h)}\cos\alpha_0\right], \qquad (8.29)$$

where n_G and n_T = the refractive indexes; at the ground and target, respectively;

r_o = the radius of the earth;

h = the target height above the earth's surface;

α_0 = the apparent elevation angle of the target.

The function $\Delta\alpha_T$ is plotted in figure 8–12 at height levels of 100,000 ft and 100 n mi, where n_T is practically zero. It is seen that $\Delta\alpha_T$ decreases monotonically with elevation angle for a given height. The region of maximum $\Delta\alpha_T$ (figure 8–13) for 0° elevation angle centers in the vicinity of

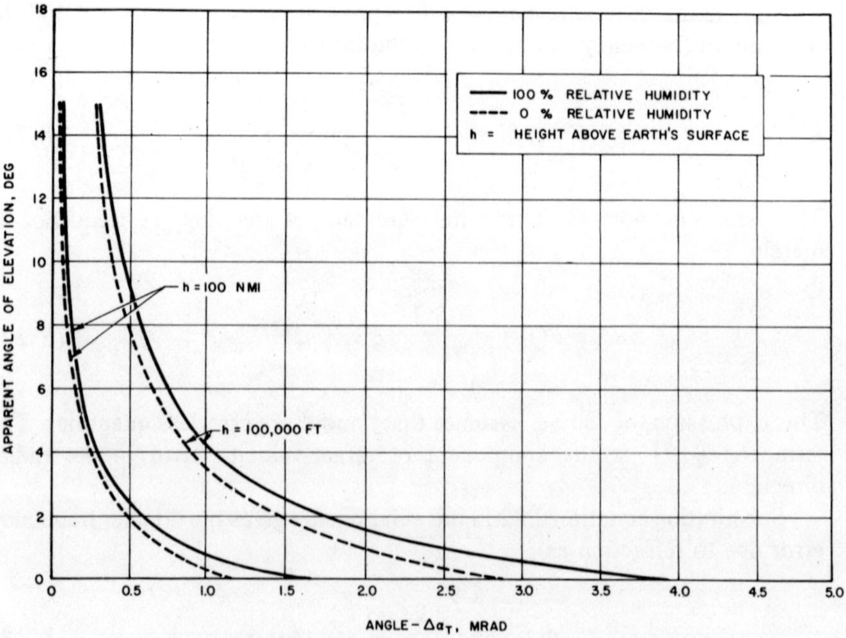

Figure 8–12. Angle $\Delta\alpha_T$ for the Troposphere

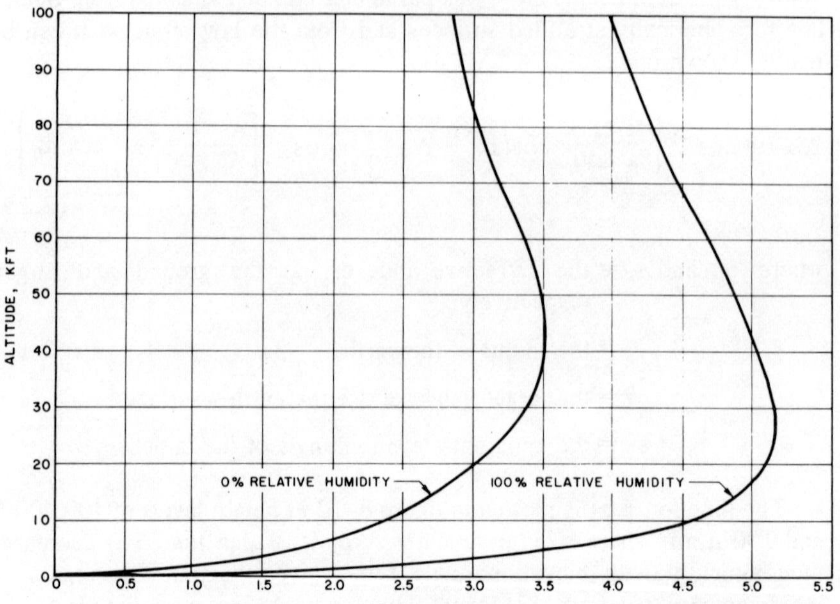

Figure 8–13. Region of Maximum $\Delta\alpha_T$ in the Troposphere for 0° Elevation

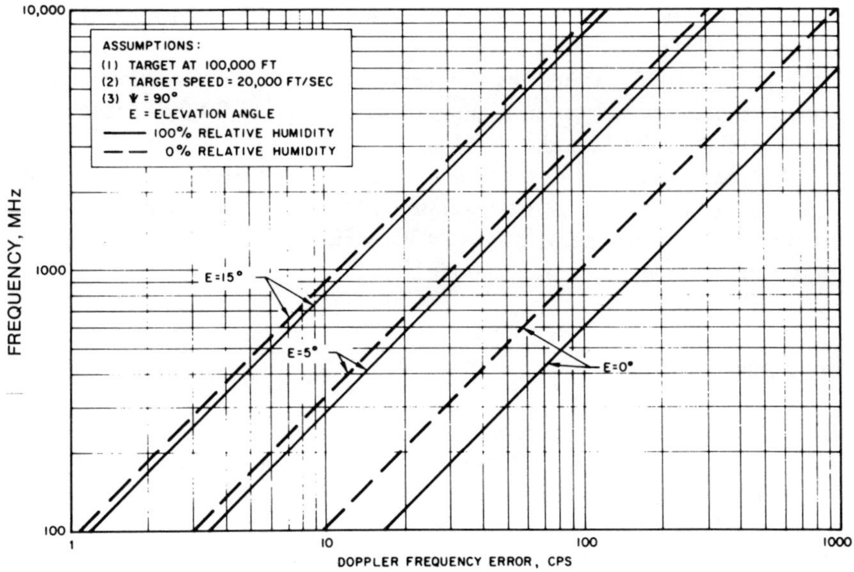

Figure 8–14. Doppler Frequency Error in the Troposphere

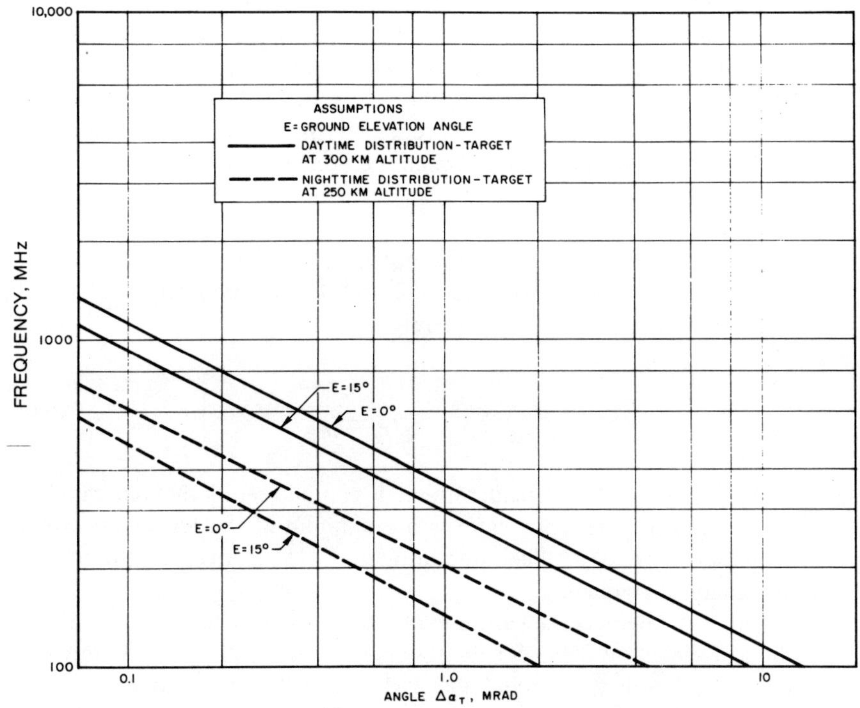

Figure 8–15. Angle $\Delta\alpha_T$ for the Ionosphere as a Function of Frequency

25,000 ft for the 100-percent relative-humidity model and 40,000 ft for the 0-percent relative-humidity model.

Figure 8–14 illustrates the doppler frequency error for the two tropospheric models based on an assumed target speed of 20,000 ft/sec, target altitude of 100,000 ft, and $\psi = 90°$.

It is seen that the greatest doppler frequency error occurs for the standard atmosphere with 100 percent relative humidity. The error is directly proportional to frequency, and at 100 MHz and 0° elevation angle the maximum error is approximately 16 Hz.

Ionospheric Effect

The mathematical derivation of the doppler frequency error just presented is valid for the ionosphere problem. However, in applying equation 8.29 to conditions in the ionosphere, the following modifications are required:

1. The radius of the earth r_o is replaced by $r_o + h_o$, where h_o is the height of the lower edge of the ionosphere above the earth's surface.
2. The target height h is measured vertically from the lower edge of the ionosphere.
3. The ground refractive index n_G now becomes n_I, the index of refraction at the entrance into the ionsophere, which for all practical purposes is unity.
4. The ground-elevation angle is replaced by the ionospheric-elevation angle, defined as the angle between the incident ionosphere ray and the tangent plane.

It can be shown that when the effects of atmospheric refraction are neglected, the relationship between the ground-elevation angle α_o and the ionospheric elevation angle α_I is merely

$$\alpha_I = \cos^{-1}\left[\frac{r_o}{r_o + h_o}\cos\alpha_o\right]. \tag{D-30}$$

From simple geometric considerations, it can be readily demonstrated that the angle $\Delta\alpha_T$ for the ionosphere is a maximum at the level of maximum electron density and, as illustrated in figure 8–15, is inversely proportional to the square of frequency.

Figure 8–16 depicts the variation of $\Delta\alpha_T$ at a frequency of 200 MHz as a function of elevation angle for an assumed target height of 300 km for the daytime-electron-density model and 250 km for the nighttime model.

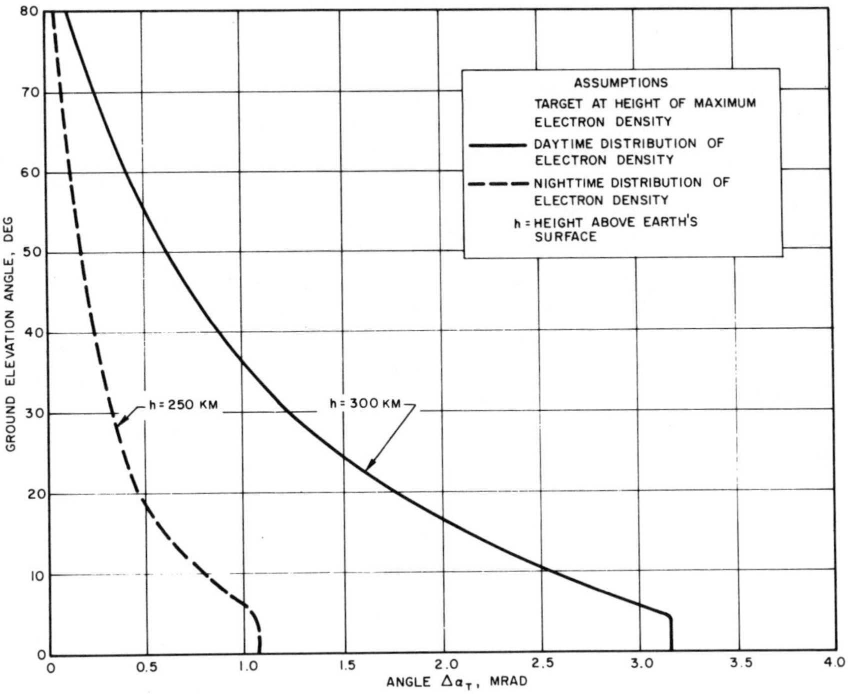

Figure 8–16. Angle $\Delta\alpha_T$ for the Ionosphere at 200 MHz

From the results presented in figure 8–17, it is seen that the doppler frequency error is at maximum during the daytime and at low elevation angles. For an assumed target speed of 20,000 ft/sec and $\psi = 90°$, the error at 100 MHz and at a ground-elevation angle of 0°, which is analogous to an ionospheric elevation angle of about 9°, is on the order of 51 Hz. For the nighttime, under similar conditions, the error is only 17 Hz. It should be mentioned that the doppler frequency error is inversely proportional to frequency for the ionosphere.

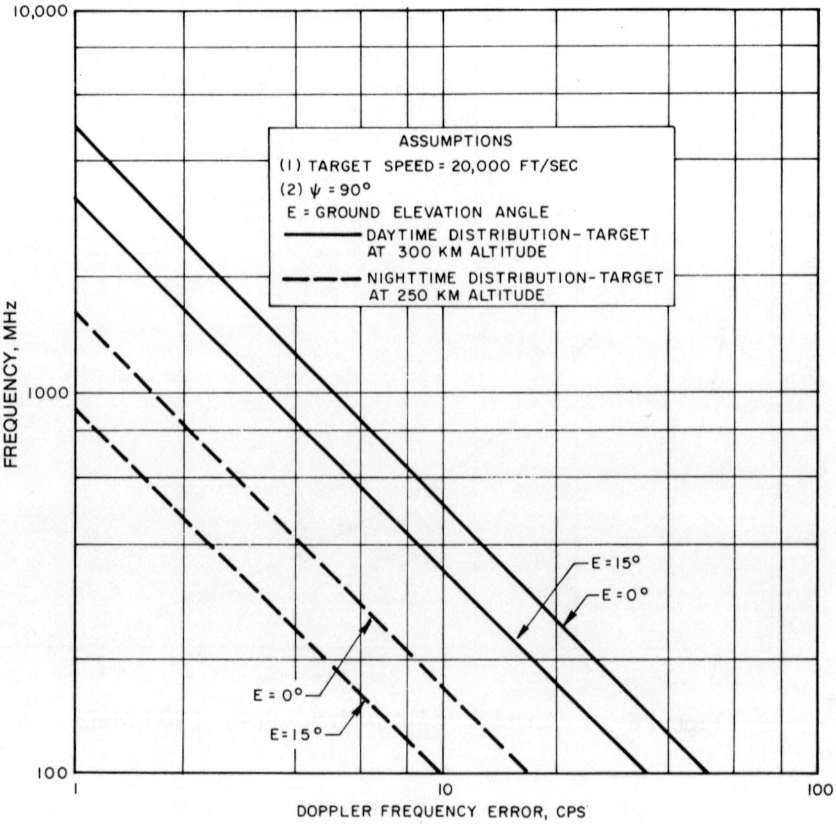

Figure 8–17. Doppler Frequency Error in the Ionosphere

9 Radar Cross-Section Models in Detection

Introduction

The approach to radar cross-section (RCS) determination and application is usually driven by the needs of the user. If a specification is to be met, the vehicle designer is primarily interested in whether the measured RCS is below (or above) a certain level, whereas the radar-systems analyst may be interested in the variation of RCS with aspect angles in the azimuthal and/or elevation plane. In the latter case, target motional effects and radar/target geometries can allow evaluation of radar performance in detection and tracking of the vehicle in the presence of natural and man-made interference. If the radar-systems analyst is interested in evaluating the efficacy of the radar in target discrimination and/or target identification, detailed radar cross-section variations as a function of frequencies, bandwidths, and polarizations may be required.

Detection and tracking of low-RCS vehicles in severe-clutter environments is perhaps the area in which statistical-detection theory is most applicable in radar performance assessment. To apply these many available techniques in a meaningful way requires that the vehicle RCS values be available in a format allowing specification of applicable statistical target models.

Discussion

Determination of the RCS of a target both by analytical means [1,2,3] and by measurement [4,5] has been developed primarily from a diagnostic and estimative point of view. The emphasis has been on identifying sources of contributions to the total signal backscattered when the target is illuminated by an electromagnetic wave, and the variation of this signal-return amplitude as a function of angle (in two orthogonal planes), frequency, and polarizations. A great deal of static-range radar cross-section data were obtained simply because this was less costly than dynamic measurement—where the target is in flight—even though the latter situation may be more realistic.

A compromise radar-analysis approach, made more attractive by the low-cost availability of the digital computer, was to combine static-range radar cross-section data with radar parameters and radar/target geometries to *simulate* the RCS time history for radar-performance evalution [6]. The result of such radar-signal simulation was to obtain a measure of the radar signature—the statistical amplitude variation of the radar cross-section as a function of time—as it would affect the radar-system parameters. From the earliest attempts to predict radar performance on a statistical basis [7], a great deal of effort has been expended on definition of useful target models in calculation of detection probabilities, and in accounting for effects of signal processing [8,9,10,20].

The direct passage from measured values of radar cross-section into statistical target models for radar-system analysis is still not clear, even though many attempts have been made to bridge this gap [11,12]. Measurements have been made of cone-cylindrical target geometries, which were not tumbling but were stable about their pitch and yaw axes, and were viewed with a radar line of sight(s) over a range of aspect angles with respect to the longitudinal (or roll) axes. Both monostatic RCS measurements at a static radar cross-section range [13] and dynamic bistatic radar cross-section measurements [14] of satellite-type targets yielded large data sets. In each case these were reduced for statistical analyses. The conclusions were similar—the so-called log-normal statistical distribution provided the optimum fit to the observed data for both bistatic and monostatic cases. Satellites in low orbit and cruise missiles both have long cylindrical shapes with vehicle stabilization about the flight path (longitudinal) axis. Their RCS variations are very similar.

The link to radar-detection probability prediction for stabilized missiles, which are physicaly similar to satellites, is first to obtain an "average" RCS value and then to use this value in the radar equation to obtain an "average" signal-to-noise ratio. One can then use available plots of detection probability for fluctuating targets [10,15] with log-normal distributions to obtain detection probabilities for various false-alarm rates and postdetection (pulse-integration) signal processing [16]. Many background-terrain and sea-clutter levels can also be described statistically by the log-normal distribution, particularly at low grazing angles. Detection probabilities for log-normal fluctuating targets in log-normal clutter backgrounds for various signal-processing methods have been documented [10,17,18,19].

Some available detection-probability curves are a function of the ratio of range R to a reference range R_o. The range R_o is obtained from radar parameters as the value of range resulting in a signal-to-noise ratio of unity, or 0 dB. Although this method is of some use in comparisons, most detection-probability curves are now given as a function of number of samples processed and signal-to-noise levels [9,10]. Several available curves display the single pulse (sample) signal-to-noise ratio with the number of pulses (samples) processed as a parameter [10,16,17]. The theory is summarized and numerical tables are given in convenient form for applications in [22].

The choice of target models and their use in computation of detection probabilities, together with a comprehensive set of convenient detectability plots, are available [10] and recommended for use in radar-performance evaluation. Refinements to the knowledge of target models for detection calculation are continuing [20], but such refinements are not expected to improve significantly the variety of methods currently available [10].

The measurement of radar cross-section for targets having a large degree of symmetry about their roll axes usually involves recording the backscatter return from nose-on (usually taken as 0°) in the azimuthal plane through broadside (near 90°) to tail-on (180°) at each degree or fraction of a degree. This is repeated at each frequency and transmit/receive polarization of interest and, if the object is not symmetrical, for the full 360° azimuth in each of several roll angles. Clearly, a great deal of RCS data can become available.

The radar-systems analyst is often interested in the RCS values near nose-on and in some contiguous range of aspect angles, depending on anticipated radar/target geometries. Long, cylindrical objects with conical noses have an RCS versus aspect angle characterized by a peak value nose-on (0°) falling off by approximately 10 dB at somewhat less than 10° from nose-on, and remaining at that level to near 60°, where the broadside specular lobe for the cylindrical body is beginning. The actual average value of RCS to be used for radar-detection prediction can vary, therefore, by approximately 10 dB, depending on whether the nose-on value or off-nose-on values are selected.

One method of obtaining an average RCS is to define the value at each of several discrete frequencies as:

$$\sigma_T(f) = \frac{1}{54}\left[\sum_{\theta_R=1}^{3}\sum_{P=1}^{3}\sum_{\theta_A=-3}^{3}\sigma_{10}(f),\right] \tag{9.1}$$

where $\sigma_T(f)$ = average target RCS at a frequency f in dB m^2;

$\sigma_{10}(f)$ = measured average value of RCS in dB m^2 in a 10° aspect-angle interval at a particular polarization P and roll angle θ_R;

θ_A = aspect-angle intervals:
−30 to −20
−20 to −10
−10 to 0
0 to +10
+10 to +20
+20 to +30, all in degrees;

θ_R = roll angles; 0°, 60°, and 90°;

P = polarization (transmit/receive) VV, HH, RCLC;

f = 2.0, 2.7, 3.6, 4.9, 6.6, 8.9, 12.0 GHz.

In this method of average-RCS computation, numerical values corresponding to the *logarithm* of the radar cross-section are averaged arithmetically, yielding a geometric mean. The definition provided by equation 9.1 is easy to calculate and does yield RCS smoothing, which is appropriate for roll-symmetric target geometries in terms of the roll angle and polarization averaging. However, if the radar-target geometry is such that the nose-on region of aspect angles is viewed (assuming no effective nose-on RCS reduction), the resulting value will be weighted in that the "average" RCS value will be less than that which determines radar-detection range. Since detection range varies as the one-fourth root of radar cross-section, all other factors being equal, the detection range could be nearly 1.7 times the values predicted by equation 9.1 if the nose-on value only were viewed by the radar, since it could be higher by a factor of 10.

The aspect-angle range for a radar located on an airborne platform of 10-km altitude, viewing an approaching target at 300-km range, might well be limited to a few degrees from nose-on, whereas a ground-based radar viewing the target at 30-km range may have a radar line of sight at several tens of degrees with respect to the vehicle's flight path. The range of aspect angles for each engagement would be quite different, as would the terrain masking and clutter return, which are also dependent on radar-site and target-approach geometries. It is also to be expected that a flight path might be chosen to present the radar with the lowest probability of detecting the vehicle, if radar locations were known.

As will be shown, the concepts of obtaining an "average" RCS can vary. In the interest of providing a numerical comparison, a representative sample-data set, as shown in table 9–1, was used to calculate the values for arithmetic mean and geometric mean. The difference for this example data

Table 9–1
RCS Sample-Data Set (Amplitude Only)

N	$\Delta\theta_i \sim deg.$	$\sigma_i \sim m^2$	
1	−30 to −20	0.04	
2	−20 to −10	0.04	
3	−10 to 0	0.4	Near nose-on
4	0 to +10	0.4	
5	+10 to +20	0.04	
6	+20 to +30	0.04	

set is less than 3 dB, comparable to variations in RCS due to polarization uncertainties; it is concluded that the decibel-averaging method for defining average RCS is reasonable. The method is convenient, and other comparable unknowns exist in application of the radar equation and detection-probability theory to prediction of radar-system performance and evaluation. RCS modeling for bistatic radar detection and at higher RCS levels may yield much larger differences and remain to be investigated.

1. Arithmetic mean calculation:

$$\bar{\sigma}_A = \frac{1}{N}\left(\sum_{i=1}^{N} \sigma_i\right), \quad \bar{\sigma}_A = 0.16 \text{ m}^2,$$

$$(\bar{\sigma}_A)_{\text{dB m}^2} = -7.96 \text{ dB m}^2.$$

2. Geometric mean calculation:

$$\bar{\sigma}_{AF} = 0.086 \text{ m}^2,$$

$$\bar{\sigma}_{AF} = \left(\prod_{i=1}^{N} \sigma_i\right)^{1/N} \quad (\bar{\sigma}_{AF})_{\text{dB m}^2} = -10.6 \text{ dB m}^2.$$

3. Difference is ≈ 2.64 dB.

RCS-Averaging Concepts

The arithmetic average of values of radar cross-section for targets of rotational symmetry (bodies of revolution), having no variation of RCS in a plane orthogonal to the longitudinal axis, can be formulated for a plane containing the longitudinal axis as

$$\bar{\sigma}_A = \frac{1}{|\theta_2 - \theta_1|} \int_{\theta_1}^{\theta_2} \sigma(\theta) d\theta, \tag{9.2}$$

where $\bar{\sigma}_A$ = arithmetic mean, or average;

$(\theta_2 - \theta_1)$ = aspect angle interval measurement, $(\theta_2 - \theta_1) \geq 0$;

$\sigma(\theta)$ = variation of RCS as a function of aspect angle.

Equation 9.1 can be written in discrete form as

$$\bar{\sigma}_A = \frac{1}{N}\left(\sum_{i=1}^{N} \sigma_i,\right) \tag{9.3}$$

where the σ_i are N samples of average RCS values in $\Delta\theta_i$ uniform aspect-angle increments, located within the aspect-angle interval under consideration.

A weighted average may be defined in a manner similar to that of equation 9.2 as

$$\bar{\sigma}_{AW} = \frac{1}{|\theta_2 - \theta_1|} \int_{\theta_1}^{\theta_2} \omega(\theta)\sigma(\theta)d\theta, \tag{9.4}$$

where the weighting function $\omega(\theta)$ provides a relative weighting over the aspect angles of interest [2]. For example, it might be desirable to weight most heavily the aspect angle subinterval most likely to be observed in a particular radar/target geometry, and to place less emphasis on those aspect angles less likely to be observed. In discrete form, equation 9.4 would become

$$\bar{\sigma}_{AW} = \frac{1}{N}\left(\sum_{i=1}^{N} \omega_i \sigma_i, \right) \tag{9.5}$$

where the individual weights ω_i might be normalized such that $\omega_i \leq 1$.

Another method of defining an average is to use a functional form [2], such as

$$\bar{\sigma}_{AF} = \frac{1}{|\theta_2 - \theta_1|} \int_{\theta_1}^{\theta_2} F\{\sigma(\theta)\}d\theta. \tag{9.6}$$

The most convenient functional form has been to use a logarithmic function—the RCS values in decibels referred to a one-square-meter value. This form is obtained as

$$(\bar{\sigma}_{AF})_{\text{dB m}^2} = \frac{1}{|\theta_2 - \theta_1|} \int_{\theta_1}^{\theta_2} 10 \log_{10} \sigma(\theta)d\theta. \tag{9.7}$$

For convenience the factor K may be used to relate the decibel logarithmic form (log) to the natural logarithmic form (ln) as

$$K \ln(\cdot) = 10 \log_{10}(\cdot). \tag{9.8}$$

Substituting equation 9.7 in 9.6 yields

$$K \ln(\bar{\sigma}_{AF}) = \frac{K}{|\theta_2 - \theta_1|} \int_{\theta_1}^{\theta_2} \ln \sigma(\theta) d\theta. \tag{9.9}$$

Equation 9.9 may be written in discrete form as

$$\ln(\bar{\sigma}_{AF}) = \frac{1}{N}\left(\sum_{i=1}^{N} \ln \sigma_i, \right) \tag{9.10}$$

or

$$10 \log_{10}(\bar{\sigma}_{AF}) = \frac{1}{N}\left[\sum_{i=1}^{N} (10 \log_{10} \sigma_i). \right] \tag{9.11}$$

Equation 9.11 affords great computational simplicity, since measured values of RCS are usually available in dB m^2 versus aspect angle. A series of uniform increments of aspect angle, covering the range of interest, yields a series of numerical values of dB m^2 that are simply summed and divided by the number of sample to provide an "average" RCS, also in units of dB m^2.

It is of interest to compare the "average" RCS values given by application of the definitions of equations 9.3 and 9.11, using 9.10, is rewritten as

$$\ln(\bar{\sigma}_{AF}) = \frac{1}{N} \ln\left(\prod_{i=1}^{N} \sigma_i \right), \tag{9.10}$$

which is equivalent to

$$\ln(\bar{\sigma}_{AF}) = \frac{1}{N}\left(\sum_{i=1}^{N} \ln \sigma_i, \right) \tag{9.12}$$

or

$$\ln(\bar{\sigma}_{AF}) = \ln\left[\left(\prod_{i=1}^{N} \sigma_i \right)^{1/N} \right]. \tag{9.13}$$

Equation 9.13, in numerical form, is then

$$(\bar{\sigma}_{AF}) = \left(\prod_{i=1}^{N} \sigma_i \right)^{1/N}, \qquad (9.14)$$

and the ratio of the arithmetic "average" to the logarithmic "average" is seen to be the ratio between an arithmetic mean and a geometric mean. Using equations 9.3 and 9.14 yields

$$\frac{\bar{\sigma}_A}{\bar{\sigma}_{AF}} = \frac{\frac{1}{N}\left(\sum_{i=1}^{N} \sigma_i \right)}{\left(\prod_{i=1}^{N} \sigma_i \right)^{1/N}}. \qquad (9.15)$$

For all nonnegative values of $\sigma_i \geq 0$, the geometric mean is equal to or *less* than the arithmetic mean [21], and

$$\left(\prod_{i=1}^{N} \sigma_i \right)^{1/N} \leq \frac{1}{N}\left(\sum_{i=1}^{N} \sigma_i \right). \qquad (9.16)$$

The "average" RCS value obtained by using numerical values in units of dB m^2 as samples for obtaining an arithmetic mean as shown in the numerator of equation 9.15 will be *lower* than the value that should be used to calculate average received signal-to-noise levels. Its use in determination of detection probability may yield erroneous results, in which actual detection ranges are somewhat greater than those predicted.

References

1. "Special Issue on Radar Reflectivity" *Proc. IEEE* 53, no. 8(August 1965): 769–1168.
2. Crispin, J.W., and Siegel, K.M. *Methods of Radar Cross Section Analysis.* New York: Academic Press, 1968.
3. Ruck, G.T., ed. *Radar Cross Section Handbook*, 2 vols. New York: Plenum Press, 1970.
4. *Radar Reflectivity Measurements Symposium*, RADC-TDR-64-25, vols. I and II: AD601364, AD601365, April 1964.
5. Bachman, C.G.; King, H.E.; and Hansen, R.C. "Techniques for

Measurement of Reduced Radar Cross Sections." BSD-TDR-62-300, Contract AF 04(695)-69, October 1962.

6. Mitchell, R.L. *Radar Signal Simulation.* Dedham, Mass.: Artech House, 1976.
7. Marcum, J.I., and Swerling, P. "Studies of Target Defection by Pulsed Radar." *Trans. IRE Prof. Gp. Information Theory* (*Special Monograph Issue*) IT-6, no. 2(April 1960).
8. Weinstock, W.W. "Radar Cross Section Target Models" and "Illustrative Problems in Radar Detection Analysis." In *Modern Radar: Analysis, Evaluation and System Design,* R.S. Berkowitz, ed., chaps. 5 and 6. New York: Wiley, 1965.
9. Swerling, P. "Recent Developments in Target Models for Radar Detection Analysis." *AGARD Conference Proceedings,* no. 66, AGARD-CP-66-70, "Advanced Radar Systems," AD 715 485, November 1970.
10. Meyer, D.P., and Mayer, H.A. "Radar Target Detection: Handbook of Theory and Practice." New York: Academic Press, 1973.
11. Rheinstein, J. "On the Comparison of Static and Dynamic Radar Cross-Section Measurements." Project Report PA-84 (BMRS), Contract AF 19 (628)-500. Massachusetts Institute of Technology, Lincoln Laboratory, 28 January 1965.
12. Air Force Technical Note no. BSD-TDR-62-381, "A Preliminary Investigation of Suitable Techniques for the Expression of Static Radar Cross Section Signatures." Chrysler Corporation, Report no. ADB-TN-37-62, Contract AF-(694)-25, 15 December 1962.
13. Kennedy, R.W. "The Spatial and Spectral Characteristics of Satellite-Type Targets." Technical Report AFAL-TR-66-17, March 1966.
14. Brescia, R.E., and Zirm, R.R. "The Distribution of Reflecting Cross Sections of Satellites." NRL Report 5615, 25 May 1961.
15. Swerling, P. "Probability of Detection for Some Additional Fluctuating Target Cases." Aerospace Corporation Report no. TOR-699 (9990) 14, Contract AF 04 (695)-669, March 1966.
16. Heidbreder, G.R., and Mitchell, R.L. "Detection Probabilities for Log-Normally Distributed Signals." SSD-TR-66-87, Aerospace Report no. TR-669 (9990)-6, April 1966. Also *IEEE Trans. PG-AES* AES-3, no. 1(January 1967): 5–13.
17. Kramer, J.D.R., et al. "The Detection Problem for Log-Normally Distributed Signal and Noise." MITRE Report MTP-136, Contract F 19628-71-C-002, May 1972.
18. George, S.F. "The Detection of Nonfluctuating Targets in Log-Normal Clutter." NRL Report 6796, 4 October 1968.

19. Trunk, G.V. "Non-Rayleigh Sea Clutter: Properties and Detection of Targets." NRL Report 7986, 26 June 1976.
20. Von Schlacta, K. "Remarks on Target Models for the Design of Radar Systems." *Proc. IEEE* (1975): 4440–4445.
21. Rektorys, K., ed. *Survey of Applicable Mathematics.* Cambridge, Mass.: MIT Press, 1969, p. 47.
22. Bachman, C.G. "Radar Sensor Engineering," Lexington Books, Lexington, Mass.: D.C. Heath and Company, 1982.

Index

About the Author

Christian G. Bachman, D.Sc., P.E., is the founder and president of KRYBOK Associates, a technical management consultancy located in Wilmington, Massachusetts. He has over twenty-five years' experience in engineering and management of programs in electronic-systems technology, and is the author of *Laser Radar Systems and Techniques* and *Radar Sensor Engineering.*

He received the B.S.E.E. from Oregon State University, the M.S.E.E. from the State University of New York, and the D.Sc. in engineering from California Western University. He has completed additional graduate work at the University of California at Los Angeles, Cornell University, and the University of Michigan. He is a graduate of the National Defense University's Industrial College of the Armed Forces, and a registered professional engineer in Massachusetts.